ROBOTS IN THE WORKPLACE

Robots in the Workplace

THE FUTURE OF EMPLOYMENT IN THE AI ERA

Dr. Hesham Mohamed Elsherif

ELDONUSA Publishing

Contents

1

ABOUT THE AUTHOR

Robots in the Workplace
The Future of Employment in The AI Era
By
Dr. Hesham Mohamed Elsherif

An expert in Empirical research methodology, Dr. Elsherif specializes particularly in the Qualitative approach and Action research. This specialization has not only strengthened his research endeavors but has also allowed him to contribute invaluable insights and advancements in these areas.

Over the years, Dr. Elsherif has made significant contributions to the academic world not only as a professional researcher but also as an

1

Adjunct Professor. This multifaceted role in the educational landscape has further solidified his reputation as a thought leader and pioneer.

Furthermore, Dr. Elsherif's expertise isn't confined to one region. He has served as a consultant to numerous educational institutions on an international scale, sharing best practices, innovative strategies, and his deep insights into the ever-evolving realms of management and technology.

Combining a passion for education with an unparalleled depth of knowledge, Dr. Elsherif continues to inspire, educate, and lead in both the library and academic communities.

2

PREFACE

As we stand at the precipice of a new era, where artificial intelligence and robotics are not just concepts of science fiction but tangible realities reshaping our world, "Robots in the Workplace: The Future of Employment in the AI Era" emerges as a timely exploration of this transformative landscape. This book is intended for a broad audience, encompassing professionals, academics, policymakers, and anyone curious about the future of work in an increasingly automated world.

In this comprehensive examination, we delve into the multifaceted implications of AI and robotics in the workplace. Our journey begins with a historical overview, tracing the evolution of technology in the workplace, from the earliest tools used by humankind to the sophisticated AI systems of today. This backdrop is essential for understanding the current state of workplace automation and its potential trajectory.

The heart of this book lies in its in-depth analysis of various industries, illustrating how AI and robotics are already revolutionizing fields such as manufacturing, healthcare, finance, and even creative domains. We present case studies and expert interviews, offering a kaleidoscope of perspectives on the changes unfolding in these sectors.

A significant focus is placed on the human aspect of this technological revolution. We explore the societal and ethical implications of a workforce increasingly intertwined with machines. What does it mean for job security, skill requirements, and workplace dynamics? How do we ensure that the benefits of AI and robotics are equitably distributed? These are some of the critical questions we address.

Furthermore, "Robots in the Workplace" delves into the future, presenting well-researched predictions and scenarios. We explore potential new job categories that might emerge and the skills that will be in high demand. This section aims not to alarm but to prepare and inspire readers for the opportunities that lie ahead.

The book also provides practical guidance for businesses and individuals looking to adapt to this new era. We offer strategies for companies to integrate AI and robotics ethically and effectively, and advice for individuals seeking to future-proof their careers.

Finally, we conclude with a discussion on policy implications. What roles should governments and international bodies play in regulating AI and robotics? How can they foster innovation while protecting workers' rights and interests? These questions are crucial in shaping a future where technology serves humanity, rather than the other way around.

"Robots in the Workplace: The Future of Employment in the AI Era" is more than just a book; it's a roadmap to navigate the uncharted territories of the AI-driven future. It invites readers to engage, reflect, and prepare for a world where the lines between human and machine labor are increasingly blurred. We hope this book ignites conversations, sparks ideas, and inspires action towards a future where technology and humanity coexist in harmony.

Dr. Hesham Mohamed Elsherif

3

WHO SHOULD READ THIS BOOK?

"Robots in the Workplace: The Future of Employment in the AI Era" is crafted for a diverse audience, each bringing their unique perspectives and seeking different insights into the rapidly evolving world of AI-driven employment. This book is an invaluable resource for multiple groups:

1. **Business Leaders and Managers:** For those at the helm of organizations, big or small, this book offers a deep dive into how AI and robotics are reshaping the business landscape. It provides insights into leveraging these technologies for competitive advantage, managing the transformation of the workforce, and understanding the ethical implications of AI integration.

2. **Policy Makers and Government Officials:** As AI redefines the employment landscape, policy makers will find this book a crucial guide in understanding the challenges and opportunities presented by AI and robotics. It offers a comprehensive view of the potential impacts on the job market, education requirements,

and the economy, assisting in the formulation of informed, future-focused policies.

3. **Human Resource Professionals:** This book is a must-read for HR professionals who are navigating the changing dynamics of recruitment, training, and employee management in an AI-driven world. It provides insights on upskilling strategies, maintaining workforce morale in the face of automation, and ethical considerations in AI deployment.

4. **Educators and Academic Researchers:** For those in academia, this book serves as a rich resource for understanding the intersection of AI, employment, and education. It offers a foundation for curriculum development, research, and preparing students for a workforce increasingly dominated by AI and robotics.

5. **Students and Job Seekers:** Students preparing to enter the workforce and job seekers looking to pivot their careers will find this book an invaluable guide. It sheds light on the skills and knowledge necessary to thrive in an AI-integrated job market and provides a glimpse into future career paths shaped by these technologies.

6. **Technology Enthusiasts and Innovators:** Individuals fascinated by AI and robotics will find this book a treasure trove of information on the latest trends, applications, and debates in the field. It serves as a comprehensive guide to understanding how these technologies are transforming workplaces across industries.

7. **General Readers with an Interest in Future Trends:** For the curious minds keen to understand how the future of work is shaping up, this book offers an accessible yet thorough exploration of AI's role in the workplace. It provides a balanced view of the opportunities and challenges posed by this technological revolution.

8. **Labor Union Representatives:** Union leaders and representatives will find this book critical in understanding how AI and

automation are impacting workers' rights and job security. It provides insights necessary for effective advocacy and negotiation in an AI-transformed work environment.

9. **Social Scientists and Economists:** This book is a valuable resource for social scientists and economists studying the societal and economic impacts of technological change. It offers a detailed analysis of how AI and robotics are reshaping labor markets, societal structures, and economic systems.

In essence, "Robots in the Workplace: The Future of Employment in the AI Era" is designed to be an essential read for anyone interested in understanding and preparing for the vast changes AI and robotics are bringing to the world of work. Whether you are directly involved in managing these changes, affected by them, or simply curious about the future, this book provides comprehensive insights and thought-provoking perspectives to navigate the AI era. Dr. Hesham Mohamed Elsherif

4

WHY THIS BOOK IS ESSENTIAL READING?

"Robots in the Workplace: The Future of Employment in the AI Era" is not just a timely publication; it's a crucial guide for understanding and navigating the unprecedented changes AI and robotics are bringing to the world of work. Here are several compelling reasons why this book is essential reading for a wide range of audiences:

1. **Comprehensive Overview of AI in the Workplace:** This book provides a holistic view of the integration of AI and robotics in various industries. It doesn't just focus on the technological aspects but also delves into the socio-economic and ethical implications, offering a multi-dimensional perspective on the subject.

2. **Future-Proofing Careers and Businesses:** For professionals and business leaders, understanding the impact of AI is critical for future-proofing careers and businesses. This book offers insights into emerging trends, necessary skill sets, and strategies to adapt and thrive in an AI-dominated landscape.

3. **Balanced and Informed Perspectives:** In a field often polarized between utopian and dystopian views, this book offers a

balanced perspective. It weighs the benefits and challenges of AI in the workplace, providing a nuanced understanding that is crucial for informed decision-making.

4. **Guidance for Policy and Decision Makers:** As AI reshapes the labor market, policy makers and decision makers will find invaluable insights in this book for developing informed strategies and regulations that balance technological advancement with the needs of the workforce.

5. **Educational Resource:** For educators and students, this book serves as a comprehensive resource, offering current and in-depth content that is crucial for academic study and research in fields related to AI, robotics, employment, and socio-economic impacts.

6. **Real-World Case Studies and Examples:** The book brings theories and predictions to life with real-world case studies and examples. These practical insights help readers visualize the actual impacts of AI in various workplace scenarios, making the content relatable and tangible.

7. **Preparation for Ethical Challenges:** As AI and robotics continue to evolve, ethical considerations become increasingly complex. This book tackles these issues head-on, preparing readers to understand and address the ethical dilemmas that accompany technological advancements in the workplace.

8. **Global Perspective:** The impact of AI and robotics on employment is a global phenomenon. This book offers a broad perspective, including studies and examples from various countries and regions, making it relevant to a global audience.

9. **Future Trends and Predictions:** For those curious about what the future holds, this book provides well-researched predictions and trends, offering a glimpse into the future of work in an AI-driven world. It helps readers anticipate and prepare for upcoming changes.

10. **Empowerment Through Knowledge:** Ultimately, this book empowers its readers by providing them with knowledge and understanding. In an era where AI and robotics are rapidly transforming the workplace, being informed is key to adapting and thriving.

In conclusion, "Robots in the Workplace: The Future of Employment in the AI Era" is essential reading because it provides a comprehensive, balanced, and insightful look at one of the most significant transformations in the world of work. It's a guide, a resource, and a roadmap for anyone seeking to understand and navigate the challenges and opportunities presented by the rise of AI and robotics in the workplace.

Dr. Hesham Mohamed Elsherif

5

Introduction

Setting the Stage: The Rise of AI in the Workplace:

In the ever-evolving narrative of human progress, few chapters are as compelling as the current one – the rise of Artificial Intelligence (AI) in the workplace. This introduction sets the stage for a profound exploration of AI's integration into the work environment, its implications, and the reshaping of what we know as 'work'.

The Dawn of a New Era

Our journey begins at the dawn of this transformative era. Here, we trace the origins of AI, from theoretical concepts and early computational experiments to the sophisticated algorithms and learning machines that pervade today's workplaces. This section illustrates how the blend of increased computational power, advanced algorithms, and vast data has propelled AI from the fringes of research labs to the forefront of business operations.

Understanding AI and Robotics

To fully grasp the implications of AI in the workplace, we must first understand what AI and robotics entail. We demystify these technologies, explaining how they work, their different types, and their capabilities. This foundational knowledge is crucial for appreciating the subsequent discussions on AI's workplace applications and impacts.

AI's Entry into Various Industries

The narrative then shifts to how various sectors have adopted AI, transforming traditional practices. We examine case studies from manufacturing, where robots and AI systems have revolutionized production lines, to service industries where AI-driven customer service bots and analytical tools are setting new standards. Each example serves to highlight the diverse applications of AI and robotics across different domains.

The Catalyst for Change

AI's integration into the workplace is not just a technological shift but a catalyst for broader change. We discuss how AI is redefining job roles, skill requirements, and even organizational structures. This section delves into the heart of the AI revolution, exploring how it influences day-to-day work, reshapes career paths, and necessitates new forms of collaboration between humans and machines.

Embracing the Change

As AI becomes more embedded in our work lives, understanding and adapting to it becomes imperative. This introduction emphasizes the need for a proactive approach – for workers to embrace lifelong learning, for organizations to foster AI-savvy cultures, and for policymakers to consider the broader societal implications.

Setting the Tone for a Thoughtful Exploration

This introduction is not just a primer on AI's rise in the workplace but sets the tone for a thoughtful exploration of what lies ahead. It poses critical questions and invites readers to ponder the future of employment in an era where AI is not just a tool, but a coworker, a collaborator, and a catalyst for unprecedented change.

As we embark on this journey, "Setting the Stage: The Rise of AI in the Workplace" aims to inform, inspire, and invoke a sense of curiosity and preparedness for the changes AI is bringing to our professional lives. The stage is set, and the story of AI in the workplace is unfolding – one of the most exciting narratives of our time.

The Key Questions and Debates on AI-Driven Employment:

As we delve deeper into the era where Artificial Intelligence (AI) reshapes the landscape of work, a myriad of questions and debates surface, challenging our understanding and expectations of employment. This section of the book, "The Key Questions and Debates on AI-Driven Employment," aims to dissect these crucial issues, offering insights into the complex interplay between AI advancements and the workforce.

Understanding the Core Questions

At the heart of our exploration are several key questions that probe the essence of AI's impact on employment. How is AI redefining job roles and industries? What are the implications for job security and the nature of work? How does AI influence the skills that workers need to thrive? These questions form the backbone of our inquiry, guiding us through the multifaceted dimensions of AI in the workplace.

The Debate Over Job Creation and Loss

One of the most pressing debates we confront is the impact of AI on job creation and loss. While some argue that AI and automation will lead to significant job displacement, others posit that these technologies will create new job categories and opportunities. We delve into this debate, examining evidence from various industries and historical precedents of technological disruptions in the job market.

The Ethical and Societal Implications

The introduction then shifts focus to the ethical and societal implications of AI-driven employment. Concerns about bias in AI algorithms, the widening skills gap, and the potential for increased inequality are at the forefront of these discussions. We also consider the ethical responsibilities of organizations deploying AI and the role of governments in regulating AI's impact on the workforce.

The Future of Work

What does the future hold for workers in an AI-dominated landscape? This question leads us to explore scenarios and predictions, contemplating how the nature of work might evolve. We look at the potential for a more flexible, freelance-based economy, the rise of remote working facilitated by AI, and the need for continuous learning and adaptation.

Balancing Optimism and Caution

Throughout this exploration, we strive to balance optimism about AI's potential benefits with caution regarding its challenges. The introduction sets the tone for a nuanced discussion, acknowledging the transformative power of AI while also recognizing the need for thoughtful policies, ethical considerations, and proactive strategies to ensure that AI-driven employment is equitable and beneficial for all.

Invitation for Engagement

This introduction is not just a summary of the key questions and debates; it is an invitation for readers to engage with these issues critically. By presenting a comprehensive overview of the current state of AI-driven employment, we aim to equip readers with the knowledge and perspective needed to navigate this rapidly evolving landscape. The future of work is being written now, and understanding these key questions and debates is essential for anyone seeking to be an informed participant in shaping that future.

6

Chapter 1: The History of Automation

Introduction to the Chapter:

Chapter 1 of "Robots in the Workplace: The Future of Employment in the AI Era" embarks on a historical journey, tracing the evolution of automation from ancient innovations to the sophisticated AI systems of today. This exploration is essential for understanding the trajectory of technological advancements and their impact on the workforce (Brynjolfsson & McAfee, 2014).

The Early Beginnings:

The chapter opens by exploring the earliest instances of automation in human history. It references works like Mushtaq's "History of Ancient Engineering" (2019), which examines ancient civilizations, such as the Egyptians and Greeks, and their early automated inventions like water clocks and automated doors. This section highlights the historical human desire to simplify labor and enhance productivity, setting a precedent for future innovations.

The quest for automation is as old as civilization itself. Referencing Oleson's "Greek and Roman Mechanical Water-Lifting Devices" (1984), it describes how the Greeks and Romans developed complex machines, including water-lifting devices, which are considered early forms of automation. The section also draws upon Mayor's "Gods and Robots" (2018) to discuss the mythical automata of ancient Greece, such as the mechanical servants crafted by the god Hephaestus, highlighting the blend of mythology and technology.

An ancient Greek analog computer used to predict astronomical positions and eclipses. Jones' research, "The Antikythera Mechanism: The First Step towards the Future" (2000), is cited to demonstrate the complexity and sophistication of this early computer. This device exemplifies the early human endeavor to harness technology for complex calculations.

Needham's work in "Science and Civilization in China" (1986) is utilized to illustrate the advanced state of automation in ancient China, including the development of automated crossbows and water clocks. Similarly, Al-Jazari's 12th-century Book of Knowledge of Ingenious Mechanical Devices, as discussed in Hill's "The Book of Knowledge of Ingenious Mechanical Devices" (1974), introduces the reader to the Middle Eastern contributions to early automation, particularly in the field of hydraulics and robotics.

Ancient civilizations were not only focused on automating tasks for convenience or entertainment but also for practical purposes like agriculture. White's "Medieval Technology and Social Change" (1962) provides insights into early forms of agricultural automation, such as the waterwheel, which revolutionized irrigation and farming methods. This aspect underscores the long-standing human effort to apply technological solutions to fundamental challenges like food production.

Key Takeaways:

- Ancient civilizations across the globe contributed significantly to the early development of automation and robotics.
- The integration of technology in everyday life, for both practical and fantastical purposes, has been a consistent theme throughout human history.

- The ingenuity displayed in early mechanical devices and automation techniques laid the groundwork for future advancements in robotics and AI.

The Industrial Revolution: A Major Turning Point:

A substantial part of the chapter is dedicated to the Industrial Revolution, a transformative period that redefined work and society. Landes' seminal work, "The Unbound Prometheus" (2003), is cited to discuss the introduction of machines like the steam engine and the power loom, which catalyzed mass production and led to the emergence of modern factories. The social and economic ramifications of these developments, including urbanization and the formation of a new working class, are explored, drawing on Hobsbawm's "The Age of Revolution" (1996).

Mechanization and Mass Production

The onset of the Industrial Revolution marked a significant shift from manual labor to mechanization. Hounshell's "From the American System to Mass Production, 1800-1932" (1984) provides an in-depth analysis of how this transition led to the emergence of mass production techniques, particularly in the United States. The text highlights the introduction of interchangeable parts and assembly line methods, which revolutionized manufacturing processes.

The Steam Engine: Catalyst of Change

The steam engine, a hallmark of the Industrial Revolution, receives special attention. As noted in Landes' seminal work, "The Unbound Prometheus" (1969), the steam engine not only facilitated more efficient manufacturing but also had a profound impact on transportation and agriculture, effectively reshaping the entire economic landscape.

Social and Economic Implications

The social and economic implications of the Industrial Revolution are examined through Ashton's "The Industrial Revolution, 1760-1830" (1948). This source provides insights into how mechanization led to urbanization, changes in labor patterns, and the rise of new social classes, setting the context for the modern industrial workforce.

Global Impact and Spread of Industrialization

The global spread of industrialization, as detailed in Pomeranz's "The Great Divergence: China, Europe, and the Making of the Modern World Economy" (2000), is also covered. This work explores how the Industrial Revolution not only transformed Western societies but also had far-reaching effects on global economic and social structures.

Technological Innovation and Its Discontents

The chapter also addresses the less discussed aspects of the Industrial Revolution, including the Luddite movement. Sale's "Rebels Against the Future: The Luddites and Their War on the Industrial Revolution: Lessons for the Computer Age" (1995) is utilized to discuss the resistance against mechanization, highlighting the fears and disruptions caused by rapid technological change.

Key Takeaways:

- The Industrial Revolution marked a fundamental shift from hand production methods to machine-based manufacturing, altering the course of human labor and economic structures.
- The steam engine played a central role in this transformation, influencing various sectors including manufacturing, transportation, and agriculture.
- The era had significant social implications, including urbanization, the creation of new social classes, and the emergence of labor movements.
- The global impact of the Industrial Revolution set the stage for the modern economic world, influencing international relations and global trade patterns.

The 20th Century: Electronics and Computing:

Transitioning into the 20th century, the chapter highlights the evolution of automation with the advent of electronics and computing.

Key milestones such as the development of programmable computers and the rise of early robotics in manufacturing are discussed, referencing Ceruzzi's "A History of Modern Computing" (2003). This section underscores the shift from mechanical to digital automation, pivotal for the AI advancements that followed.

The Dawn of the Computer Age

The chapter highlights the transition from mechanical to electronic methods of computation, beginning with the development of early computers. Campbell-Kelly and Aspray's "Computer: A History of the Information Machine" (2004) provides a comprehensive overview of this transition, detailing pivotal inventions such as the ENIAC, the world's first general-purpose electronic computer.

The Microelectronics Revolution

The microelectronics revolution, marked by the invention of the transistor and the integrated circuit, fundamentally changed the landscape of computing and automation. Riordan and Hoddeson's "Crystal Fire: The Invention of the Transistor and the Birth of the Information Age" (1997) is cited to illustrate how these innovations led to the miniaturization and subsequent explosion of electronic devices and computers.

The Rise of Personal Computing

The evolution of personal computing is a key focus, with Ceruzzi's "A History of Modern Computing" (2003) offering insights into the development of the personal computer and its profound impact on both the workplace and society at large. This section discusses the democratization of computing technology and the beginning of the digital era.

Software Development and the Internet

The chapter also delves into the evolution of software and the rise of the internet, pivotal in shaping modern automation and AI. Abbate's "Inventing the Internet" (2000) traces the development of the internet from a government project to a global communication and information sharing tool, emphasizing its role in facilitating the growth of digital automation.

Robotics and Automation in Manufacturing

The latter part of the 20th century saw significant advancements in robotics, particularly in manufacturing. The role of robotics in transforming manufacturing processes is explored through the lens of Pistorius and Utterback's "Multi-mode Interaction Among Technologies" (1997), which examines the interaction between new and existing technologies in industrial settings.

Key Takeaways:

- The 20th century marked the advent of the computer age, transitioning from mechanical to electronic methods of computation and data processing.
- The microelectronics revolution, particularly the development of transistors and integrated circuits, played a critical role in the miniaturization and proliferation of computing devices.
- The rise of personal computing and the internet revolutionized the way individuals and businesses interact with technology, setting the foundation for today's digital world.
- Robotics and automation technologies saw significant advancements, particularly in manufacturing, fundamentally changing production processes and labor dynamics.

The Dawn of AI and Robotics

Focusing on the late 20th and early 21st centuries, the chapter delves into the emergence of AI and robotics. It discusses the development of machine learning and the growing integration of AI across various industries, supported by Russell and Norvig's "Artificial Intelligence: A Modern Approach" (2016). The evolution of robotics from simple automated machines to complex systems capable of diverse tasks is examined, drawing on insights from Brooks' "Robot: The Future of Flesh and Machines" (2002).

The Conceptual Foundations of AI

The journey into AI begins with its conceptual and theoretical foundations. Russell and Norvig's seminal work, "Artificial Intelligence: A Modern Approach" (2016), offers an extensive overview of the early ideas that shaped AI, including the Turing Test and the concept of machine learning. This part of the chapter discusses how these foundational concepts set the stage for the development of AI as we know it today.

Early Robotics and Automation

The evolution of robotics in the context of automation is examined through the lens of Siciliano and Khatib's "Springer Handbook of Robotics" (2016). This source delves into the development of the first industrial robots, such as Unimate, and their initial applications in manufacturing. It highlights the technological innovations that enabled these early robots to perform specific, repetitive tasks, revolutionizing production lines.

AI in the Digital Age

Kurzweil's "The Age of Intelligent Machines" (1992) provides insights into how the digital revolution spurred the growth of AI. This reference discusses the transition from theoretical AI to practical applications, marking the beginning of AI's integration into various industries and everyday life.

The Integration of AI and Robotics

The synergy between AI and robotics is a critical aspect of this section. Brooks' work, "Robot: The Future of Flesh and Machines" (2002), is used to discuss how advancements in AI have been integrated into robotic technology, leading to the creation of more autonomous, intelligent machines. This part of the chapter underscores the significance of this integration in shaping the current landscape of robotics and AI.

Ethical and Philosophical Considerations

The early ethical and philosophical questions surrounding AI and robotics are also addressed. Bostrom's "Superintelligence: Paths, Dangers, Strategies" (2014) is cited to explore the potential implications and ethical dilemmas posed by the advancement of AI and robotics, setting the stage for ongoing debates in the field.

Key Takeaways:

- The foundational concepts of AI, including the Turing Test and machine learning, laid the groundwork for the development of modern AI.
- The emergence of industrial robots marked a significant milestone in automation, revolutionizing manufacturing processes.
- The digital age facilitated the rapid growth of AI, leading to its integration into various sectors and everyday life.
- The integration of AI with robotics has led to the creation of more autonomous and intelligent machines, reshaping the landscape of both fields.
- Early developments in AI and robotics raised important ethical and philosophical questions, many of which continue to be relevant in current discussions.

Looking Back to Look Forward:

In its conclusion, Chapter 1 reflects on the historical lessons from the evolution of automation. It discusses how technological advancements have historically disrupted and created jobs, altered economies, and shifted societal structures, with insights from Autor's "Why Are There Still So Many Jobs? The History and Future of Workplace Automation" (2015).

Learning from History

Tetlock and Gardner's "Superforecasting: The Art and Science of Prediction" (2015) provides a framework for how historical analysis can enhance our ability to forecast future developments. This source is used to underline the significance of examining past technological advancements to anticipate future changes in AI and robotics.

Industrial Revolutions and Their Lessons

The narrative revisits the industrial revolutions, drawing lessons from their impacts on employment and society. Landes' "The Unbound Prometheus: Technological Change and Industrial Development in Western Europe from 1750 to the Present" (2003) is referenced to illustrate how each industrial revolution brought about significant shifts in labor markets and economic structures, suggesting parallels with the ongoing AI revolution.

Technological Unemployment: Historical Perspectives

The concept of technological unemployment is explored with references to historical instances where technology displaced jobs. Autor's "Why Are There Still So Many Jobs? The History and Future of Workplace Automation" (2015), published in the *Journal of Economic Perspectives*, provides an analysis of how economies adapted to technological changes in the past, offering insights into potential future scenarios in the AI-driven employment landscape.

The Evolution of Work and Skills

The section also examines the evolution of work and skills in response to technological advancements. Brynjolfsson and McAfee's "The Second Machine Age: Work, Progress, and Prosperity in a Time of Brilliant Technologies" (2014) is used to discuss how the nature of work and required skills have continually adapted to technological changes, hinting at what might be needed in an AI-dominated future.

Preparing for the Future

The chapter is emphasizing the importance of learning from history to prepare for future challenges and opportunities presented by AI and robotics. Schwab's "The Fourth Industrial Revolution" (2017) is cited to argue for proactive strategies in education, policy-making, and business to harness the potential of AI and robotics while mitigating their disruptive effects.

Key Takeaways

- Historical analysis is crucial for forecasting and preparing for future technological trends and their impact on employment.

- Lessons from past industrial revolutions can provide valuable insights into how the AI revolution might reshape the labor market and economy.
- Understanding the historical context of technological unemployment can inform our responses to future job displacements due to AI and robotics.
- The evolution of work and skills in response to technological changes highlights the need for adaptability and lifelong learning in the AI era.
- Proactive strategies in various sectors are essential to harness the benefits and mitigate the challenges of the ongoing technological revolution.

The industrial revolutions and their impacts on jobs:

The First Industrial Revolution:

The first industrial revolution, originating in the late 18th century, marked the transition from hand production methods to machines and introduced steam power and factory systems. Hobsbawm's "Industry and Empire: The Birth of the Industrial Revolution" (1999) offers a detailed analysis of this period, highlighting how innovations like the spinning jenny and steam engine transformed textile manufacturing and other industries. This source underscores the initial displacement of artisan skills but also notes the creation of new job opportunities in emerging factory settings.

The Second Industrial Revolution:

Transitioning to the second industrial revolution in the late 19th and early 20th centuries, the focus shifts to mass production and technological advancements like electricity and the assembly line. Landes' "The Unbound Prometheus: Technological Change and Industrial Development in Western Europe from 1750 to the Present" (2003) provides a comprehensive view of this era, illustrating how these innovations led to a surge in production capabilities and a fundamental shift in labor dynamics, with new roles emerging in manufacturing and management.

The Third Industrial Revolution:

The third industrial revolution, or the digital revolution, began in the mid-20th century and is characterized by the advent of electronics, telecommunications, and computers. Rifkin's "The Third Industrial Revolution: How Lateral Power is Transforming Energy, the Economy, and the World" (2011), details how this era brought about the automation of production processes and the rise of the service sector, significantly altering the employment landscape with the introduction of information technology jobs.

The Fourth Industrial Revolution:

The chapter then addresses the ongoing fourth industrial revolution, marked by the fusion of technologies blurring the lines between the physical, digital, and biological spheres. Schwab's "The Fourth Industrial Revolution" (2017) is utilized to discuss the integration of AI, robotics, the Internet of Things (IoT), and other digital technologies, emphasizing their transformative impact on jobs – not just in terms of automation but also in creating new roles and industries.

Reflections and Implications for the Future:

In summarizing the impacts of these revolutions, the chapter reflects on the patterns of job displacement and creation, skill shifts, and the evolving nature of work. Autor's "Why Are There Still So Many Jobs? The History and Future of Workplace Automation" (2015), found in the *Journal of Economic Perspectives*, is referenced to argue that, like past industrial revolutions, the current era of AI and automation may lead to significant job transformations rather than outright eliminations.

Early fears and outcomes: Lessons from history:

The Luddite Movement:

The early 19th century saw the rise of the Luddite movement, a group of English workers who destroyed machinery, particularly in the textile industry, fearing job losses and reduced wages. Sale's "Rebels Against the Future: The Luddites and Their War on the Industrial Revolution: Lessons for the Computer Age" (1996) provides an in-depth look at this movement, presenting it as one of the first significant instances of technology-induced job anxiety. However, as the industrial revolution progressed, it became evident that while some jobs were lost, many more were created in new sectors, thus offsetting the initial job displacement.

The Automation Anxiety in the 20th Century:

During the 1950s and 1960s, automation anxiety became prevalent as computers and automated machinery began to enter various industries. Cohen and Zysman's "Manufacturing Matters: The Myth of the Post-Industrial Economy" (1987) discusses the fears of massive unemployment due to automation. These fears, however, were largely unfounded as the economy adjusted, and new job categories emerged, particularly in the service sector and technology-related fields.

The Role of Economic and Policy Responses:

Understanding the role of economic and policy responses in mitigating these fears is crucial. Autor, Levy, and Murnane's study (2003), presented in "The Skill Content of Recent Technological Change: An Empirical Exploration" in the *Quarterly Journal of Economics*, argues that policy interventions, education, and training played a significant role in aiding workers to transition to new job types, thus alleviating some of the initial apprehensions.

Comparing Past and Present Fears:

The chapter then draws parallels between past fears and the current apprehensions regarding AI and robotics. Brynjolfsson and McAfee's "The Second Machine Age: Work, Progress, and Prosperity in a Time of Brilliant Technologies" (2014) is referenced to illustrate how, like in the past, current fears might be overemphasized. The authors argue that while AI and robotics will undoubtedly cause disruptions, they also hold the potential for creating new job categories and enhancing productivity.

The evolution of workplace automation:

The Genesis of Mechanical Automation:

The journey begins with the early mechanical inventions of the Industrial Revolution. Hounshell's "From the American System to Mass Production, 1800-1932" (1984) provides an insightful analysis of the initial phase of automation, marked by inventions like the steam engine and mechanized looms. These innovations shifted the nature of work from manual labor to machine operation, leading to the rise of factories and mass production.

Electrification and Assembly Lines:

The early 20th century witnessed further transformation with the advent of electrification and the assembly line, revolutionizing manufacturing processes. Hughes's "Networks of Power: Electrification in Western Society, 1880-1930" (1983) explores how these technologies increased production efficiency and introduced new jobs in machine maintenance and assembly line work, yet also led to the deskilling of certain labor sectors.

The Computer Age:

The mid-20th century marked the onset of the Computer Age, fundamentally changing the automation landscape. Campbell-Kelly and Aspray's "Computer: A History of the Information Machine" (2004) traces the development of early computers and their initial application in business and industry, emphasizing the shift towards data processing and management jobs.

The Rise of Robotics and AI:

The late 20th and early 21st centuries saw the emergence of robotics and artificial intelligence, as described in Ford's "Rise of the Robots: Technology and the Threat of a Jobless Future" (2015). This era introduced robots capable of performing complex tasks, AI systems that could process vast amounts of data, and machine learning algorithms enhancing decision-making processes in businesses. These advancements led to a significant transformation in industries such as manufacturing, logistics, and even services, creating new job roles while also rendering some obsolete.

The Current State of Automation:

The current state of workplace automation is characterized by an ever-increasing integration of AI and robotics into various sectors. Schwab's "The Fourth Industrial Revolution" (2017) provides a contemporary view of how these technologies are shaping the present and future of work, highlighting the growing need for skills in technology management, data analysis, and AI ethics.

7

Chapter 2: AI's Immediate Impact

This portion of the book delves into how AI is currently reshaping various aspects of work, from task automation to decision-making processes.

AI in Task Automation:

The application of AI in automating routine tasks has been a game-changer in many industries. As Bessen (2019) articulates in "AI and Jobs: The Role of Demand," the immediate impact is most evident in sectors like manufacturing, where AI-driven robots perform tasks with precision and efficiency. This automation not only enhances productivity but also shifts the nature of human jobs towards more supervisory and maintenance roles.

The integration of Artificial Intelligence (AI) in automating tasks has been one of the most profound developments in the modern workplace. This section delves deeper into how AI-driven automation is transforming various industries and job roles.

Manufacturing Sector:

In the manufacturing sector, AI's role in automating tasks has been particularly significant. Ford (2015) in "Rise of the Robots: Technology and the Threat of a Jobless Future," discusses how AI and robotics have revolutionized production lines. Robots, equipped with AI, perform tasks ranging from assembly to quality control, leading to increased efficiency and productivity. However, this shift has also raised concerns about job displacement in routine manufacturing roles.

Service Industry:

In the service industry, task automation through AI has been transformative. Davenport and Kirby (2016), in their work "Only Humans Need Apply: Winners and Losers in the Age of Smart Machines," examine how AI tools like automated scheduling and customer service chatbots are reshaping service delivery. While these tools enhance operational efficiency, they also necessitate a workforce skilled in managing and integrating these technologies.

Healthcare:

In healthcare, AI's impact in task automation is evident in diagnostic processes and patient care. Topol (2019) in "Deep Medicine: How Artificial Intelligence Can Make Healthcare Human Again," highlights how AI algorithms assist in analyzing medical images and patient data, aiding in more accurate diagnoses and personalized treatment plans. This advancement underscores the need for healthcare professionals to adapt to a tech-integrated work environment.

Administrative Tasks:

AI's reach extends to automating administrative tasks as well. Kaplan and Haenlein (2019), in "Siri, Siri, in my hand: Who's the fairest in the land? On the interpretations, illustrations, and implications of artificial intelligence," note the efficiency gains in sectors like finance and legal, where AI-driven software can handle data entry, scheduling, and basic research tasks, freeing human workers for more complex activities.

Decision Making and AI:

AI's role in decision-making processes has been transformative, particularly in data-driven fields. Agrawal, Gans, and Goldfarb (2018) in "Prediction Machines: The Simple Economics of Artificial Intelligence" highlight how AI algorithms are now integral in analyzing large data sets, enabling more informed decisions in areas like finance, healthcare, and logistics. This shift demands a new skill set from workers, focusing on data interpretation and strategic thinking.

The incorporation of Artificial Intelligence (AI) in decision-making processes marks a significant shift in various industries. This section explores in greater detail how AI is influencing and enhancing decision-making capabilities in different sectors.

Business and Finance:

In the realm of business and finance, AI's role in decision making is profoundly impactful. Bughin, Hazan, and Ramaswamy (2017) in their work "Artificial Intelligence: The Next Digital Frontier?" discuss how AI algorithms analyze market trends and consumer data to inform business strategies and financial decisions. AI tools in finance, as noted by Arner, Barberis, and Buckley (2017) in "FinTech and RegTech in a Nutshell," are reshaping investment strategies and risk management through predictive analytics, offering insights that were previously unattainable.

Healthcare:

The healthcare sector benefits immensely from AI in decision making, particularly in patient care and treatment plans. Topol (2019) in "Deep Medicine: How Artificial Intelligence Can Make Healthcare Human Again" illustrates how AI algorithms interpret patient data and medical research to assist healthcare professionals in making more accurate and personalized treatment decisions. This AI-assisted decision-making is revolutionizing patient outcomes and the efficiency of healthcare systems.

Public Policy:

In public policy, AI's influence is growing in areas like urban planning and social welfare programs. Tegmark (2017) in "Life 3.0: Being Human in the Age of Artificial Intelligence," highlights how AI systems are used to analyze social trends, environmental data, and economic indicators to assist policymakers in making informed decisions that affect communities and societies.

Retail and E-commerce:

In retail and e-commerce, AI's role in decision making is evident through personalized marketing and inventory management. Marr (2016) in "Data Strategy: How to Profit from a World of Big Data, Analytics and the Internet of Things," details how AI algorithms track consumer behavior to tailor marketing strategies and manage stock levels, leading to more efficient and customer-centric business models.

AI in Customer Service:

The adoption of AI in customer service, as discussed by Huang and Rust (2018) in "Artificial Intelligence in Service," shows how AI technologies like chatbots and virtual assistants are revolutionizing customer interactions. This not only leads to efficiency gains but also requires employees to develop skills in managing and integrating AI tools into the customer experience.

The integration of Artificial Intelligence (AI) in customer service is transforming how businesses interact with and serve their customers. This expansion delves into the specifics of AI's role in revolutionizing customer service across different industries.

Personalized Customer Interactions:

AI has made it possible to offer highly personalized customer service experiences. According to Huang and Rust (2018) in "Artificial Intelligence in Service," AI systems, through data analysis and machine learning, can tailor interactions based on customer preferences and history, thereby enhancing customer satisfaction. This personalization is not just limited to recommendations but extends to the entire customer journey, as discussed by Meuter, Bitner, Ostrom, and Brown (2005) in "Choosing Among Alternative Service Delivery Modes: An Investigation of Customer Trial of Self-Service Technologies."

Chatbots and Virtual Assistants:

One of the most visible impacts of AI in customer service is the proliferation of chatbots and virtual assistants. Gnewuch, Morana, Maedche, and Weinhardt (2017), in their study "Towards Designing Cooperative and Social Conversational Agents for Customer Service," highlight how these AI-powered tools provide instant, round-the-clock assistance to customers. These tools handle routine inquiries efficiently, freeing human agents to tackle more complex issues, as observed by Feine, Morana, and Maedche (2019) in "A Taxonomy of Social Cues for Conversational Agents."

Handling Large Volume of Customer Queries:

AI systems are adept at managing large volumes of customer queries, which is particularly beneficial for businesses during peak periods or special events. A study by Davenport, Guha, Grewal, and Bressgott (2020) in "How Artificial Intelligence Will Change the Future of Marketing" outlines how AI can quickly analyze and respond to customer inquiries, maintaining high levels of service even under pressure.

Enhancing Customer Support Quality:

AI's role in enhancing the quality of customer support is significant. Kaplan and Haenlein (2019) in "Siri, Siri, in my hand: Who's the fairest in the land? On the interpretations, illustrations, and implications of artificial intelligence," discuss how AI-powered tools are capable of learning and improving over time, leading to continuously improving service quality. This continuous learning capability ensures that customer service operations become more efficient and effective.

The Changing Skill Landscape:

The immediate impact of AI is also evident in the changing skill landscape. As Manyika et al. (2017) in "Jobs Lost, Jobs Gained: Workforce Transitions in a Time of Automation" discuss, there's a growing need for digital literacy, complex problem-solving abilities, and an understanding of AI and machine learning concepts among the workforce.

The rapid integration of Artificial Intelligence (AI) in various sectors is drastically altering the skill landscape in the job market. This expanded discussion examines how AI is reshaping the required skills and job roles in the contemporary workplace.

Emergence of New Skill Sets:

The advent of AI necessitates new skill sets, particularly in data analysis, machine learning, and AI management. Bessen (2019), in "The Business of AI Startups," emphasizes the growing demand for skills in AI programming and data science. Moreover, skills in AI ethics and governance are becoming increasingly important, as discussed by Hagendorff (2020) in "The Ethics of AI Ethics: An Evaluation of Guidelines."

Upgrading Traditional Skills:

AI's influence extends to traditional job roles, where there is a need for upgrading skills to work alongside AI systems. As Wilson and Daugherty (2018) argue in "Collaborative Intelligence: Humans and AI Are Joining Forces," there is a growing need for workers to develop skills that complement AI, such as decision-making and problem-solving skills. This transition is not limited to tech roles but spans across various sectors, including healthcare, finance, and manufacturing.

Continuous Learning and Adaptation:

The dynamic nature of AI technology necessitates continuous learning and adaptation. This aspect is highlighted by Schwab (2016) in "The Fourth Industrial Revolution," where he discusses the need for ongoing education and training to keep pace with technological advancements. Lifelong learning becomes crucial, as seen in the increasing relevance of online courses and training programs focused on AI and related technologies.

The Role of Educational Institutions:

Educational institutions are pivotal in preparing the workforce for this transition. A study by Luckin et al. (2016) in "Intelligence Unleashed: An Argument for AI in Education" explores how educational systems are adapting curricula to include AI and data science, preparing students for the AI-driven job market.

Ethical Considerations and AI:

Moreover, AI's immediate impact raises crucial ethical considerations. Bostrom and Yudkowsky (2014) in "The Ethics of Artificial Intelligence" explore the ethical challenges posed by AI, including privacy concerns, algorithmic bias, and the need for transparent AI systems. These considerations are becoming increasingly important in workplaces where AI is used for decision-making.

The integration of Artificial Intelligence (AI) into various aspects of the workplace brings with it a host of ethical considerations. This expanded discussion delves into the multifaceted ethical challenges posed by AI and the efforts to address them.

AI and Privacy Concerns:

One of the most pressing ethical issues is the management of privacy in the age of AI. As AI systems often require vast amounts of data to function effectively, this raises concerns about data security and privacy. The work of Richards and King (2014), in their article "Big Data Ethics," outlines the challenges in balancing data utility and privacy. They stress the importance of implementing robust privacy policies and ethical guidelines in the handling of personal data.

Bias and Fairness in AI Systems:

Another significant ethical issue is the potential for bias in AI systems. This can manifest in various forms, from gender and racial biases to socio-economic biases. Barocas, Hardt, and Narayanan (2019) in their book "Fairness and Abstraction in Sociotechnical Systems" highlight the challenges in ensuring AI systems are fair and unbiased. They argue for the development of AI systems that are transparent and accountable in their decision-making processes.

AI and Employment Ethics:

The impact of AI on employment also raises ethical questions. The displacement of jobs by AI and automation can have significant social and economic implications. A study by Acemoglu and Restrepo (2020), in "Robots and Jobs: Evidence from US Labor Markets," examines the effects of automation on employment and wages. They emphasize the need for policies that mitigate the negative impact on workers and promote equitable economic growth.

Ethical AI Governance:

The governance of AI is crucial in addressing these ethical concerns. Jobin, Ienca, and Vayena (2019) in their study "The Global Landscape of AI Ethics Guidelines" review various AI ethics guidelines proposed around the world. They note the convergence on principles like transparency, fairness, and accountability but also the differences in implementation strategies.

Current industries undergoing AI transformation:

Overview of Industry-Specific AI Transformations:

Artificial Intelligence (AI) is revolutionizing various industries, each facing unique challenges and opportunities. This section examines key industries undergoing significant transformations due to AI integration, supported by relevant literature.

Healthcare: AI-Driven Diagnostics and Treatment

In healthcare, AI's impact is profound, particularly in diagnostics and treatment planning. Jiang et al. (2017) discussed AI applications in healthcare, emphasizing its role in enhancing diagnostics accuracy and personalizing treatment. AI algorithms can analyze medical images or genetic information to identify diseases earlier and more accurately than traditional methods.

Artificial Intelligence (AI) is significantly reshaping healthcare, particularly in diagnostics and treatment. This section provides a deeper insight into the specific areas within healthcare that are being transformed by AI, supported by relevant academic sources.

Enhanced Diagnostic Accuracy with AI

One of the most notable applications of AI in healthcare is in improving the accuracy of diagnostics. Esteva et al. (2019) emphasized how deep learning algorithms are used in dermatology to identify skin cancer with a level of competence comparable to dermatologists. These algorithms, trained on vast datasets of skin images, can detect nuances in skin lesions that might be missed by the human eye.

AI in Genetic Sequencing and Personalized Medicine

AI's role extends to genetic sequencing and personalized medicine. Obermeyer et al. (2016) discussed how machine learning techniques are applied to genomic data to predict patient responses to different treatments. This personalized approach allows for more effective and tailored treatment plans, potentially revolutionizing cancer treatment and other genetic-based diseases.

Improving Radiology with AI

In radiology, AI is transforming image analysis. Langlotz (2019) highlighted the use of AI in interpreting medical imaging, such as X-rays, CT scans, and MRIs. AI algorithms enhance the detection of abnormalities, often spotting issues earlier and with greater accuracy than traditional methods. This advancement not only improves diagnostic accuracy but also reduces the workload on radiologists.

AI in Drug Discovery and Development

AI is also making significant strides in drug discovery and development. Zhavoronkov et al. (2019) explored how AI is utilized in the pharmaceutical industry, particularly in speeding up the drug discovery process and in predicting drug efficacy and safety. AI algorithms can analyze vast chemical and biological data sets to identify potential new drugs and predict their interactions, streamlining the development process.

Financial Services: AI in Risk Assessment and Fraud Detection

The financial sector is leveraging AI for risk assessment and fraud detection. Bholat et al. (2018) explored how AI is used in finance, particularly in enhancing the accuracy of risk models and detecting fraudulent activities. AI algorithms analyze patterns in financial transactions to identify anomalies that may indicate fraudulent behavior.

The integration of Artificial Intelligence (AI) in financial services, particularly in risk assessment and fraud detection, represents a significant shift in how financial institutions manage risk and security. This section delves deeper into these areas, supported by pertinent academic sources.

AI in Enhancing Risk Assessment

In the context of risk assessment, AI algorithms offer sophisticated tools for analyzing large datasets to identify potential risks and make informed decisions. Bhatia et al. (2019) discussed how AI can process and analyze various types of data, including transaction history, market trends, and customer behavior, to assess creditworthiness more accurately than traditional models. This capacity for nuanced analysis enables financial institutions to make more informed lending decisions, reducing the likelihood of defaults.

AI in Fraud Detection

AI's role in fraud detection is particularly noteworthy. Van Vlasselaer et al. (2015) explored how machine learning algorithms are increasingly used to detect unusual patterns indicative of fraudulent activities. By analyzing transaction data in real time, these algorithms can identify anomalies that might elude traditional detection methods, thereby enhancing the security of financial transactions.

Predictive Analytics in Financial Services

Furthermore, AI is integral in predictive analytics within financial services. Khandani et al. (2010) emphasized the use of AI in predicting market trends and customer behavior, which is crucial for risk management. By forecasting potential market shifts and customer actions, financial institutions can proactively manage risks associated with market volatility and customer creditworthiness.

Retail: Personalization and Inventory Management

In retail, AI is transforming customer experiences and inventory management. Huang and Rust (2018) discussed how AI is changing service delivery in retail, focusing on personalized shopping experiences and efficient inventory management. AI systems analyze consumer behavior to tailor product recommendations and optimize stock levels.

The retail sector has seen a notable transformation with the integration of Artificial Intelligence (AI), especially in the domains of customer personalization and inventory management. This section explores these applications in greater detail, citing relevant academic sources.

AI in Retail Personalization

AI-driven personalization in retail has revolutionized the shopping experience. According to Grewal et al. (2020), AI algorithms analyze customer data, including past purchases, browsing history, and preferences, to offer tailored recommendations and personalized shopping experiences. This level of personalization not only enhances customer satisfaction but also increases sales by suggesting products that customers are more likely to purchase.

AI in Inventory Management

In terms of inventory management, AI plays a crucial role in predicting demand, optimizing stock levels, and reducing waste. Verhoef et al. (2021) highlighted how AI algorithms analyze sales data, seasonal trends, and market changes to forecast future product demand accurately. This predictive capability enables retailers to maintain optimal inventory levels, minimizing both overstock and stockouts, and thereby ensuring a more efficient supply chain.

AI-Driven Customer Engagement

Additionally, AI significantly contributes to enhancing customer engagement in retail. Huang and Rust (2018) examined how AI technologies, like chatbots and virtual assistants, provide personalized customer service, improving customer engagement and loyalty. These AI tools can handle a range of customer inquiries, provide shopping assistance, and even offer personalized product recommendations, thereby enriching the overall customer experience.

In summary, AI's impact in the retail sector is multi-faceted, significantly enhancing both customer personalization and inventory management. Through sophisticated data analysis and predictive capabilities, AI helps retailers tailor their offerings to individual customer preferences while simultaneously managing inventory in a more efficient and cost-effective manner. These advancements not only improve the customer experience but also streamline operational processes, showcasing the transformative power of AI in the retail industry.

Manufacturing: AI in Predictive Maintenance and Supply Chain Optimization

Manufacturing is seeing significant benefits from AI, especially in predictive maintenance and supply chain optimization. Lee, Kao, and Yang (2014) highlighted AI's role in predictive maintenance, where AI algorithms predict equipment failures before they occur, reducing downtime. AI also optimizes supply chains by analyzing patterns and predicting demand.

The manufacturing industry is undergoing a significant transformation with the integration of Artificial Intelligence (AI), particularly in predictive maintenance and supply chain optimization. This section delves deeper into these applications, supported by academic sources.

AI in Predictive Maintenance

Predictive maintenance, powered by AI, marks a significant shift from traditional reactive maintenance strategies. AI algorithms analyze data from sensors and machines to predict equipment failures before they occur. According to Lee et al. (2019), this AI-driven approach enhances the efficiency of maintenance tasks, reduces downtime, and saves costs associated with unexpected equipment failures. By leveraging machine learning and data analytics, manufacturers can schedule maintenance activities more effectively, ensuring optimal equipment performance and longevity.

AI in Supply Chain Optimization

AI's role in optimizing supply chains in manufacturing is equally transformative. A study by Ivanov and Dolgui (2020) highlights how AI algorithms provide dynamic supply chain planning and management. These systems analyze vast amounts of data, including supplier performance, material availability, and demand forecasts, to optimize production schedules, inventory levels, and distribution routes. This capability not only ensures a more responsive and flexible supply chain but also helps in mitigating risks associated with supply chain disruptions.

Enhancing Manufacturing Efficiency

Furthermore, AI significantly contributes to overall manufacturing efficiency. As noted by Zhou et al. (2018), AI technologies facilitate the automation of routine tasks, improve quality control with advanced vision systems, and enhance decision-making processes. This leads to increased productivity, reduced operational costs, and improved product quality.

In summary, AI's role in the manufacturing sector is vital, particularly in predictive maintenance and supply chain optimization. By analyzing data and predicting outcomes, AI enables manufacturers to preemptively address potential issues and optimize their operations. This not only enhances efficiency and productivity but also contributes to more robust and resilient supply chains. The adoption of AI in manufacturing signifies a move towards more intelligent, efficient, and responsive industrial processes.

Job roles augmented and replaced by AI:

The integration of AI in various sectors is leading to significant changes in job roles, with certain positions being augmented or entirely replaced by AI technologies. This section explores these trends in depth, supported by academic sources.

AI-Augmented Roles

AI augmentation refers to the enhancement of human capabilities and decision-making processes through AI technologies. According to a study by Daugherty and Wilson (2018), AI augmentation is more about assisting and improving human skills rather than replacing them. For example, in healthcare, AI algorithms assist doctors in diagnosing diseases more accurately and quickly, thus augmenting their capabilities. In fields like finance, AI assists analysts in processing vast amounts of data to make more informed decisions.

AI's role in augmenting jobs is a critical aspect of its impact on the workforce. This enhancement of human capabilities through AI is reshaping numerous professions, enabling more efficient and accurate outcomes.

Healthcare

In the healthcare sector, AI augmentation is revolutionizing diagnostics and treatment. Jiang et al. (2017) discuss how AI algorithms are being used for early detection of diseases like cancer, often with greater accuracy than traditional methods. This not only assists healthcare professionals in making more informed decisions but also improves patient outcomes. AI tools, such as image recognition systems, are aiding radiologists in identifying anomalies that might be missed by the human eye.

Finance

In finance, AI's impact is predominantly seen in data analysis and risk assessment. AI algorithms can process vast datasets to identify trends and insights, aiding financial analysts in making more informed decisions. Brougham and Haar (2018) emphasize how AI tools have become indispensable in areas like investment strategy formulation and fraud detection, enhancing the efficiency and effectiveness of financial professionals.

Education

The education sector is also experiencing AI-driven augmentation. AI tools are being used to create personalized learning experiences, as discussed by Luckin et al. (2016). These technologies adapt to individual learning styles and pace, enabling educators to provide more targeted and effective instruction.

AI augmentation is thus not just about replacing human effort but enhancing it, leading to more efficient, accurate, and personalized outcomes in various sectors. This trend highlights the increasing importance of AI literacy and adaptability in the workforce, as professionals must learn to work alongside and leverage AI technologies effectively.

The synergistic relationship between humans and AI is key to unlocking the full potential of this technological advancement.

Roles Replaced by AI

On the other hand, certain job roles are being replaced by AI, especially those involving repetitive and routine tasks. A seminal paper by Frey and Osborne (2017) suggests that jobs in transportation, logistics, and basic customer service are highly susceptible to automation through AI. For instance, AI-driven chatbots and automated customer service systems are increasingly handling tasks that were once the domain of human customer service representatives.

The advancement of AI has led to a significant transformation in the job market, with certain roles being completely replaced by AI technologies. This shift reflects the increasing capabilities of AI in handling tasks that were traditionally performed by humans.

Manufacturing

In the manufacturing industry, the advent of AI and robotics has led to the automation of various labor-intensive tasks. According to Acemoglu and Restrepo (2020), repetitive and predictable tasks, such as assembly line work, have been largely taken over by AI-driven machines. This shift has resulted in increased efficiency and production but also raised concerns about the displacement of unskilled labor.

Data Entry and Analysis

AI has also replaced roles in data entry and analysis. AI algorithms are capable of processing vast amounts of data more quickly and accurately than humans. Smith and Anderson (2014) discuss how AI systems in sectors like banking and insurance are being used to automate data entry tasks, reducing the need for manual data entry clerks.

Customer Service

Automated customer service agents, powered by AI, are increasingly handling customer inquiries and complaints. These AI systems can manage a high volume of requests simultaneously, providing 24/7 service that human employees cannot. Kaplan and Haenlein (2019) note that these systems are becoming more sophisticated, able to handle a wider range of queries with human-like interaction.

The replacement of certain job roles by AI is a double-edged sword. While it brings efficiency and cost-effectiveness to businesses, it also poses challenges for the workforce, particularly those in roles that are susceptible to automation. This phenomenon underscores the need for policy interventions and educational reforms to prepare the current and future workforce for an AI-dominated job market. Retraining and upskilling programs are vital to ensure that workers are not left behind in the rapidly evolving employment landscape. The transition to an AI-driven workplace demands a balanced approach, where the benefits of AI are harnessed while mitigating its disruptive effects on employment.

AI's Impact on Job Market Dynamics

The impact of AI on employment is complex and multifaceted. As highlighted by Arntz, Gregory, and Zierahn (2016), while AI may lead to the displacement of certain jobs, it also creates new roles and opportunities, particularly in AI development, maintenance, and oversight. The dynamic nature of this transition necessitates a focus on re-skilling and up-skilling of the workforce to adapt to these changes.

The integration of Artificial Intelligence (AI) into various industries has fundamentally altered the dynamics of the job market. This transformation is characterized not only by the augmentation and replacement of certain roles but also by the creation of new job categories and the reshaping of career paths.

Creation of New Job Categories

AI has led to the emergence of new job categories that require specialized skills in AI and machine learning. Bessen (2019) highlights how the tech industry has seen a surge in demand for AI specialists, data scientists, and machine learning engineers. These roles focus on developing, managing, and optimizing AI systems and are pivotal in guiding AI applications across sectors.

Reshaping Career Paths

The integration of AI in the workplace is reshaping traditional career paths. As AI takes over routine tasks, the focus shifts to roles that require complex problem-solving, creativity, and interpersonal skills. Wilson and Daugherty (2018) emphasize that employees are now required to adapt and acquire new skills to work alongside AI technologies effectively.

Displacement and Job Transition

AI's ability to automate tasks has led to displacement in certain sectors, primarily affecting low-skill, repetitive jobs. Frey and Osborne (2017) argue that this displacement presents challenges in terms of job transitions and re-skilling. Workers in affected industries must seek new career paths, often requiring additional training and education.

AI as a Job Market Equalizer

Despite concerns of job displacement, AI has the potential to act as a market equalizer. Bughin et al. (2018) suggest that AI can create more inclusive work environments by enabling remote work and flexible schedules, thus making it easier for a broader range of individuals to participate in the workforce, including those with disabilities or care-giving responsibilities.

The impact of AI on job market dynamics is multifaceted. While it presents challenges in terms of displacement and the need for re-skilling, it also offers opportunities for career development in emerging fields and the potential for a more inclusive workforce. Understanding and adapting to these changes is crucial for both employees and employers to navigate the evolving landscape of employment in the AI era. Continuing education and policy interventions will play a significant role in facilitating this transition and ensuring that the benefits of AI in the workplace are realized across the spectrum of the job market.

Early economic and societal impacts:

Economic Impacts

The early economic impacts of AI are diverse and significant, affecting productivity, business models, and market dynamics.

1. **Productivity Gains**:

AI has the potential to significantly boost productivity. According to a report by Accenture (2016), AI could double annual economic growth rates by 2035 through changing the nature of work and creating a new relationship between man and machine. The productivity gains are attributed to AI's ability to process and analyze vast amounts of data more efficiently than humans.

The introduction of AI into various sectors has led to significant productivity gains, primarily through its ability to optimize operations, enhance decision-making, and innovate in product and service delivery.

1. **Optimization of Operations**: AI excels in streamlining and optimizing operational processes. According to a study by Bughin, Hazan, and Ramaswamy (2017), AI can improve production efficiency by analyzing data patterns to identify optimization opportunities. This leads to reduced operational costs and increased throughput.

2. **Enhanced Decision-Making**: AI's capacity for data analysis surpasses human capabilities, especially in handling large volumes of data and identifying trends. McKinsey & Company (2018) observed that AI-driven data analysis helps businesses make informed decisions quickly, leading to more efficient resource allocation and better business outcomes.

3. **Innovation in Product and Service Delivery**: AI is not just a tool for improving existing processes; it also drives innovation. Davenport and Ronanki (2018) discuss how AI has enabled the creation of new products and services, particularly in industries

like finance, healthcare, and retail, where personalized customer experiences are increasingly important.

4. **Impact on Labor Productivity**: The effects of AI on labor productivity are profound. According to a report by the World Economic Forum (2018), AI and automation are expected to lead to a net increase in jobs and productivity. However, this increase is contingent upon the workforce's ability to adapt and the creation of new roles that AI and automation cannot fulfill.

5. **Sector-Specific Productivity Improvements**: Different sectors experience varying degrees of productivity gains from AI. For example, Agrawal, Gans, and Goldfarb (2017) highlight that in sectors like manufacturing, where processes can be extensively automated, productivity gains are significant. In contrast, in more creative industries, AI acts more as an augmenting tool rather than a direct replacement, leading to different kinds of productivity improvements.

In summary, AI's contribution to productivity gains is multifaceted, enhancing not only operational efficiency but also decision-making processes and innovation across various industries. While these advancements promise significant economic benefits, they also require adaptive strategies in workforce development and business operations to fully realize their potential.

II. Transformation of Business Models:

AI is reshaping business models across various industries. Bughin, Chui, and Manyika (2018) note that AI enables new ways of performing tasks, creating products, and interacting with customers, leading to the emergence of new business models and revenue streams.

The advent of AI technologies has catalyzed a transformation in business models across industries. This shift is characterized by new ways of value creation, customer engagement, and revenue generation.

1. **Value Creation through Data**: AI's ability to process and analyze large datasets has turned data into a critical asset for businesses. As Bughin, Seong, Manyika, Chui, and Joshi (2018) illustrate, companies now leverage AI to extract insights from data, which they use to create more value for their customers, often leading to entirely new product offerings or services.

2. **Customer Engagement and Personalization**: AI has revolutionized customer engagement by enabling personalized experiences. Huang and Rust (2018) discuss how AI technologies, like chatbots and recommendation systems, allow businesses to interact with customers in a more tailored manner, enhancing customer satisfaction and loyalty.

3. **New Revenue Streams**: The integration of AI has opened up novel revenue streams. A study by Taddy (2019) highlights that AI can identify new market opportunities and customer needs, enabling businesses to tap into previously unexplored revenue sources.

4. **Shift in Competitive Dynamics**: The deployment of AI reshapes competitive dynamics within industries. According to a report by McKinsey & Company (2019), companies that effectively harness AI capabilities gain a competitive edge, as they can operate more efficiently and innovate faster than their competitors.

5. **Changing Nature of Work and Employment**: AI-induced transformations extend to the workforce. As Kaplan and Haenlein (2019) note, AI has changed the nature of work by automating routine tasks and demanding new skill sets from employees, leading to a shift in job roles and functions within organizations.

6. **AI-Driven Business Agility**: The flexibility and adaptability of businesses have been enhanced by AI. Davenport, Guha, Grewal, and Bressgott (2020) argue that AI empowers businesses to respond quickly to market changes and customer demands, thus increasing their agility and resilience.

In essence, AI is not just a technological upgrade but a catalyst for fundamental changes in how businesses operate, compete, and deliver value. It necessitates a strategic rethinking of business models to harness its potential fully, impacting both the internal dynamics of organizations and their external market interactions.

III. Market Dynamics and Competition:

The implementation of AI is altering market dynamics and competitive landscapes. As observed by Brynjolfsson and McAfee (2017), AI-driven companies are gaining significant advantages over competitors, leading to market shifts and increased pressures on traditional businesses to innovate.

Artificial Intelligence (AI) has profoundly influenced market dynamics and competition, reshaping industries and altering the competitive landscape. This impact is multifaceted, affecting everything from market structures to competitive strategies.

1. **Market Structure Transformation**: AI is a significant force in transforming market structures. As Agrawal, Gans, and Goldfarb (2019) discuss, AI is creating markets that are more data-driven and efficient, often leading to the emergence of new market leaders who excel in AI capabilities.

2. **Competition Intensification**: The integration of AI in business processes has intensified competition across industries. As observed by Porter and Heppelmann (2018), companies leveraging AI effectively can achieve superior performance, forcing competitors to innovate rapidly or risk obsolescence.

3. **Barriers to Entry and Market Concentration**: AI can simultaneously lower and raise market entry barriers. For instance, Brynjolfsson and McAfee (2017) highlight that while AI technologies can be accessible to new entrants, the significant data and expertise required to train AI systems effectively can favor established players, potentially leading to increased market concentration.

4. **Disruptive Innovation and Business Models**: AI enables disruptive innovation, altering traditional business models. As described by Christensen, Raynor, and McDonald (2015), AI-driven innovations often disrupt existing markets by offering new types of value propositions, leading to the creation or destruction of entire industry segments.

5. **Strategic Decision Making**: The impact of AI on strategic decision-making is profound. Davenport and Ronanki (2018) argue that AI technologies provide businesses with enhanced insights and predictive capabilities, allowing for more informed and strategic decisions.

6. **Consumer Behavior and Expectations**: AI has also influenced consumer behavior and expectations. According to Huang and Rust (2018), AI-driven personalization and recommendation systems are shaping consumer preferences, demanding businesses to adapt their strategies accordingly.

In summary, AI's impact on market dynamics and competition is complex and far-reaching. It not only reshapes how companies compete but also changes the very nature of markets and consumer expectations, thereby requiring a strategic reevaluation for businesses to stay competitive in the AI era.

Societal Impacts

AI's societal impacts are equally profound, influencing employment, ethical considerations, and societal well-being.

1. **Employment Shifts:**

AI's role in automating tasks has led to shifts in employment patterns. Frey and Osborne (2017) discuss how AI is likely to displace a range of jobs, particularly those involving routine tasks, while simultaneously creating new roles in tech and AI development.

The initiation of Artificial Intelligence (AI) has instigated significant shifts in employment patterns, impacting various sectors of society. This shift encompasses changes in job types, the emergence of new roles, and the displacement of certain jobs.

1. **Job Type Transformation**: AI has led to a transformation in the nature of jobs. As Susskind and Susskind (2015) point out, tasks that involve pattern recognition and data processing are increasingly being automated, altering the job landscape. This shift is creating a demand for skills that AI cannot replicate easily, such as creativity, emotional intelligence, and complex problem-solving.

2. **Emergence of New Roles**: With the growth of AI, new job roles are emerging. Bughin et al. (2018) in their McKinsey report highlight the creation of positions such as AI trainers, maintainers, and managers, which are essential for the effective implementation and oversight of AI systems.

3. **Displacement of Traditional Jobs**: AI's impact on employment is not uniformly positive. Frey and Osborne (2017) have famously estimated that up to 47% of current jobs in the United States are at risk of being automated. This displacement is particularly pronounced in sectors like manufacturing, transportation, and administrative support.

4. **Reskilling and Upskilling Needs**: The transition to an AI-driven economy necessitates significant reskilling and upskilling of the workforce. The World Economic Forum (2018) emphasizes the need for continuous learning and adaptation, as the skills required in the workplace evolve rapidly.

5. **Inequality and Social Stratification**: The differential impact of AI on various job segments can exacerbate social inequalities. Autor (2015) discusses how AI may lead to greater income disparities, as high-skill workers benefit more from AI advancements than low-skill workers.

6. **Shifts in Labor Markets**: The labor market itself is undergoing changes due to AI. Kaplan and Haenlein (2019) suggest that AI is contributing to a more fluid, gig-based economy, where traditional full-time employment may become less common.

Essentially, the societal impacts of AI on employment are multifaceted, encompassing both opportunities and challenges. While AI drives innovation and creates new job types, it also poses significant risks of job displacement and necessitates a rethinking of skills development and employment strategies in the AI era.

II. Ethical Considerations:

The deployment of AI raises important ethical issues, including concerns about privacy, surveillance, and decision-making biases. Crawford and Calo (2016) highlight the need for frameworks to address these ethical challenges, ensuring AI is developed and used responsibly.

The integration of Artificial Intelligence (AI) into various sectors has raised significant ethical considerations. These concerns extend beyond the immediate economic impacts to broader societal implications.

1. **Bias and Fairness**: AI systems, reflecting the biases present in their training data, can perpetuate and amplify societal biases. Noble (2018) in her book "Algorithms of Oppression" highlights how algorithms can reinforce racial and gender biases, leading to discriminatory outcomes in areas like job recruitment, healthcare, and law enforcement.

2. **Privacy and Surveillance**: The use of AI in surveillance and data analysis poses serious privacy concerns. Zuboff (2019) discusses the concept of "surveillance capitalism," where personal data is commodified by tech companies, often without users' explicit consent, challenging traditional notions of privacy.

3. **Accountability and Transparency**: As AI systems become more complex, the issue of accountability arises. Bostrom and Yudkowsky (2014) address the "black box" problem in AI, where

decision-making processes are not transparent, making it difficult to attribute responsibility for AI's actions or decisions.

4. **Autonomy and Human Dignity**: AI's role in decision-making raises questions about human autonomy. Brynjolfsson and McAfee (2014) discuss how AI-driven decisions can undermine human dignity, especially when these decisions are about critical life aspects such as employment, healthcare, and criminal justice.

5. **Long-term Societal Changes**: The long-term societal impact of AI, including changes in social structures and human behavior, is a growing concern. Harari (2016) speculates on the future of societies shaped by AI, suggesting potential challenges to democracy, freedom, and what it means to be human.

6. **AI in Warfare**: The use of AI in military applications, including autonomous weapons, raises ethical dilemmas. Sharkey (2010) warns of the moral and legal challenges posed by autonomous weapon systems, which could change the nature of warfare and accountability in armed conflicts.

In summary, the ethical considerations surrounding AI's immediate impact are profound and multifaceted. They encompass issues of bias, privacy, accountability, human autonomy, societal change, and the ethical use of AI in warfare. Addressing these concerns requires a collaborative effort from technologists, ethicists, policymakers, and the wider public to ensure the responsible development and deployment of AI technologies.

III. Impact on Societal Well-being:

AI's influence extends to broader societal well-being. Bostrom (2014) raises concerns about the long-term existential risks associated with advanced AI, emphasizing the need for careful stewardship of AI development.

The dawn of Artificial Intelligence (AI) significantly influences societal well-being, encompassing both positive and negative aspects. The impact of AI on societal well-being is multifaceted, involving psychological, social, and ethical dimensions.

1. **Mental Health and Psychological Impact**: AI applications in social media and digital platforms have been linked to mental health concerns. Twenge and Campbell (2018) highlight how excessive use of social media, powered by AI algorithms that promote engagement, correlates with increased rates of depression and anxiety, particularly among adolescents.

2. **Social Connectivity and Isolation**: While AI enables greater connectivity through social media and communication platforms, it also poses risks of social isolation. Turkle (2017) argues that AI-driven interactions, despite providing instant connectivity, often lack the emotional depth and fulfilment derived from human interactions, potentially leading to feelings of loneliness and isolation.

3. **Impact on Cultural and Social Norms**: AI influences cultural and social norms, particularly through media and entertainment. Jenkins (2006) explores how AI-driven personalization in media can create "filter bubbles," limiting exposure to diverse perspectives and potentially leading to cultural polarization.

4. **Ethical and Moral Development**: The ubiquity of AI in daily life raises questions about its influence on ethical and moral development. Greene et al. (2019) discuss how reliance on AI for decision-making might affect individuals' ability to make ethical judgments, potentially leading to a reliance on algorithmic solutions for moral dilemmas.

5. **Human-Robot Interactions and Empathy**: The increasing use of AI in caregiving and companionship roles, especially in elderly care, raises concerns about the impact on human empathy. Sparrow and Sparrow (2006) caution that over-reliance on AI for emotional support could diminish human empathy and the intrinsic value of human relationships.

6. **Digital Inequality and Accessibility**: The digital divide is exacerbated by AI, where access to AI technologies is unevenly distributed. Van Dijk (2020) discusses how this digital inequality affects societal well-being, with marginalized communities often being left behind in the benefits AI offers.

The impact of AI on societal well-being is complex and multifaceted, involving psychological, social, cultural, ethical, and accessibility considerations. While AI has the potential to enhance societal well-being in numerous ways, it also poses significant challenges that must be addressed to ensure equitable and positive outcomes for all members of society.

The early impacts of AI on the economy and society are far-reaching and complex. While AI drives productivity and innovation, it also necessitates consideration of ethical implications and societal well-being. It's imperative for policymakers, business leaders, and the broader community to engage in proactive dialogue and policy-making to harness AI's potential while mitigating its risks. This includes investing in education and training to prepare the workforce for AI-induced changes and establishing ethical guidelines for AI development and deployment.

8

———

Chapter 3: Job Displacement: Reality vs. Myth

The debate surrounding job displacement due to Artificial Intelligence (AI) oscillates between alarmist predictions and optimistic reassurances. Understanding the reality versus myth in this context involves examining empirical data and theoretical insights.

Extent of Job Displacement:

Many fear that AI will lead to massive job displacement. Frey and Osborne (2017) estimate that up to 47% of U.S. jobs are at risk of automation. However, this perspective often overlooks the potential for job creation and transformation. Bessen (2019) argues that technology historically creates more jobs than it destroys, suggesting a similar trend with AI.

The extent of job displacement due to Artificial Intelligence (AI) is a topic of considerable debate, with various studies presenting differing perspectives. This expansion delves deeper into the nuances of this issue.

1. **Predictions of Displacement**: The study by Frey and Osborne (2017) is often cited for its prediction that 47% of U.S. jobs are at risk of automation. This projection is based on the assumption of technological capabilities and the economic feasibility of implementing AI in various sectors. However, this figure is sometimes misinterpreted as an inevitable outcome rather than a theoretical upper limit under specific conditions.

2. **Differentiating Between Task and Job Automation**: A key distinction in understanding job displacement is between automating tasks and automating entire jobs. Many jobs consist of a variety of tasks, some of which are more susceptible to automation than others. Arntz, Gregory, and Zierahn (2016) argue that when considering the heterogeneity of tasks within jobs, the proportion of jobs at risk drops significantly, to around 9% in OECD countries.

3. **Sector-Specific Impacts**: The extent of job displacement varies significantly across sectors. Sectors with high routine and predictable tasks, such as manufacturing and data processing, are more prone to automation. In contrast, sectors that require human interaction and creative problem-solving, like healthcare and education, are less susceptible (Manyika et al., 2017).

4. **Geographical Variations**: The impact of AI on job displacement is not uniform across different regions. Studies show that areas with a higher concentration of industries more susceptible to automation face a greater risk of job losses (Autor & Dorn, 2013). This geographical factor adds complexity to the overall understanding of job displacement due to AI.

5. **Economic Growth and Job Creation**: While automation can lead to job displacement, it also contributes to economic growth, which can create new jobs. Bessen (2019) highlights that historical patterns of technological advancement have often led to net job creation, though the transition can be challenging for those whose jobs are displaced.

6. **The Role of Policy and Economic Decisions**: The actual extent of job displacement is also influenced by policy and economic decisions. Governments and industries can mitigate risks through strategic investments in education, training, and job creation in emerging sectors (Acemoglu & Restrepo, 2020).

In summary, the extent of job displacement due to AI is influenced by a complex interplay of technological capabilities, the nature of tasks within jobs, sector-specific factors, geographical variations, and the overarching economic and policy landscape. This multifaceted view challenges the notion of widespread job displacement as an inevitable outcome of AI advancement.

Nature of Jobs Affected:

The assumption that AI will only impact low-skill jobs is a myth. Many high-skill professions, including those in healthcare and law, are also susceptible to AI automation (Susskind & Susskind, 2015). The key factor is not the skill level but the nature of tasks – repetitive versus creative and interactive tasks.

The nature of jobs affected by Artificial Intelligence (AI) is a critical aspect of understanding the broader impact of technological advancements on the workforce. This expanded analysis delves into the characteristics of jobs most susceptible to AI-driven displacement and transformation.

1. **Routine vs. Non-Routine Tasks**: A fundamental determinant of a job's susceptibility to AI is the nature of its tasks. Jobs consisting predominantly of routine tasks, which are repetitive and predictable, are more likely to be automated. Non-routine tasks, which require problem-solving, creativity, and emotional intelligence, are less prone to automation (Autor, Levy, & Murnane, 2003).

2. **Cognitive vs. Manual Jobs**: Another dimension to consider is whether the job involves cognitive or manual skills. While initial

waves of automation impacted manual jobs in sectors like manufacturing, recent advancements in AI and machine learning are increasingly affecting cognitive jobs, particularly those involving data processing and analysis (Brynjolfsson & McAfee, 2014).

3. **Skills and Education Level**: Generally, jobs requiring lower levels of education and training are at higher risk of automation. Jobs that require higher education, specialized skills, or advanced training tend to involve complex tasks that are more challenging to automate (Autor, 2015).

4. **Industry-Specific Variations**: The susceptibility of jobs to AI varies significantly across industries. For instance, jobs in transportation and warehousing are highly susceptible due to the routine nature of tasks and the rapid development of autonomous vehicle technology. In contrast, jobs in healthcare and education are less at risk due to the high degree of human interaction and ethical considerations involved (Manyika et al., 2017).

5. **Socio-Economic Factors**: The impact of AI on jobs is also mediated by socio-economic factors. Workers in lower-income brackets, who often occupy jobs with a high degree of routine tasks, are disproportionately affected by automation. This leads to concerns about increasing income inequality and job polarization (Acemoglu & Restrepo, 2020).

6. **Job Transformation vs. Replacement**: It's important to differentiate between job transformation and job replacement. While some jobs may be entirely replaced by AI, many others will transform, requiring workers to adapt to new roles and skills. This transformation often leads to the augmentation of human capabilities rather than outright displacement (Bessen, 2019).

In conclusion, the nature of jobs affected by AI is shaped by the type of tasks involved, the skill level required, industry-specific factors, and socio-economic contexts. While automation poses risks for certain jobs, particularly those involving routine tasks, it also creates opportunities for job transformation and the emergence of new roles. Understanding these dynamics is crucial for policymakers, educators, and industry leaders in navigating the transition towards an increasingly AI-integrated workforce.

Job Transformation vs. Replacement:

While some jobs may be fully automated, many will transform with AI integration. Autor, Levy, and Murnane (2003) demonstrate that technology complements complex tasks, enhancing job roles rather than replacing them entirely. The focus is increasingly on how AI can augment human skills.

The discourse on job displacement due to Artificial Intelligence (AI) often centers around the dichotomy of job transformation versus job replacement. This expanded analysis explores these concepts in greater depth, drawing on recent academic and industry insights.

1. **Job Transformation**: Job transformation refers to the evolution of existing jobs, wherein certain tasks are automated while new tasks emerge. This evolution often leads to the augmentation of human work rather than its diminution. AI tools can enhance productivity, allowing employees to focus on more complex and creative tasks. Bessen (2019) highlights that technological change historically has led to job transformation, where new technologies complement human skills, leading to increased demand for these jobs.

2. **Job Replacement**: In contrast, job replacement involves the outright substitution of human labor with AI or automation technologies. This is more likely in roles where the majority of tasks can be codified and automated. Frey and Osborne (2017) estimated that up to 47% of US employment is at high risk of automation, particularly in sectors like transportation, logistics, and administrative support.

3. **Skill Requirement Shifts**: With the transformation and replacement dynamics, there is an observable shift in the skill

requirements of the workforce. Skills that are uniquely human, such as emotional intelligence, creativity, and complex problem-solving, are becoming more valuable. World Economic Forum (2020) reports an increasing demand for higher cognitive skills, including creativity, critical thinking, and decision-making.

4. **Economic and Sectoral Variability**: The extent of job transformation and replacement varies significantly across economies and sectors. In advanced economies with high labor costs, there's a greater incentive to replace labor with AI in sectors like manufacturing. In contrast, in sectors where human touch is integral, such as healthcare and education, transformation is more pronounced (Autor, 2015).

5. **Future Job Creation**: While AI leads to job displacement in some areas, it also contributes to the creation of new job categories and industries. Historically, technological advancements have been net creators of jobs. The emergence of AI-specific roles, such as AI trainers, ethicists, and maintenance professionals, is indicative of this trend (Manyika et al., 2017).

6. **Policy and Education Implications**: The transformation and replacement dynamics necessitate proactive policy and educational responses. Governments and educational institutions need to focus on reskilling and upskilling programs to prepare the workforce for the evolving job landscape. There is a growing emphasis on lifelong learning and adaptability as key competencies in an AI-driven economy (Schwab, 2016).

In summary, the impact of AI on the workforce is not a simple binary of job transformation versus replacement. It is a complex interplay of factors involving the type of industry, nature of tasks, skill requirements, and socio-economic conditions. While AI does pose challenges in terms of job displacement, it also offers opportunities for job creation and enhancement. Understanding and addressing these nuances is vital for stakeholders to effectively navigate and leverage the changes brought by AI in the job market.

Economic and Policy Factors:

The impact of AI on jobs is not solely a technological issue but is deeply entwined with economic and policy decisions. Acemoglu and Restrepo (2020) emphasize that choices regarding AI deployment in workplaces are influenced by economic incentives and regulatory frameworks, affecting the extent of job displacement.

The discussion on AI-induced job displacement is not complete without a thorough understanding of the intertwined economic and policy factors. This expanded analysis delves into these aspects, supported by relevant literature.

1. **Economic Factors Influencing Job Displacement**:

- **Market Dynamics**: The pace and nature of AI adoption are significantly influenced by market dynamics, including competition, consumer demand, and technological advancements. Firms in highly competitive sectors are more inclined to adopt AI to gain an edge, potentially leading to greater job displacement (Acemoglu & Restrepo, 2018).
- **Cost of Implementation vs. Labor Costs**: The decision to replace human labor with AI is often a cost-benefit analysis. In regions with higher labor costs, companies might find AI adoption more economically viable, leading to higher job displacement (Brynjolfsson & McAfee, 2014).
- **Industry-Specific Economic Conditions**: Certain industries are more susceptible to automation due to their economic structure. For instance, manufacturing industries with repetitive tasks are prime candidates for automation (Autor, 2015).

1. **Policy Factors Influencing Job Displacement**:

- ○ **Government Regulations and Incentives**: Government policies play a crucial role in shaping AI adoption. Regulations concerning AI and labor, tax incentives for automation, and policies on data privacy can either accelerate or slow down AI integration in workplaces (Whittaker et al., 2018).
- ○ **Education and Training Policies**: Policies that encourage reskilling and upskilling of the workforce can mitigate the impact of job displacement. Investment in education and vocational training is essential to prepare workers for the changing job landscape (Schwab, 2016).
- ○ **Social Safety Nets**: The presence and robustness of social safety nets, including unemployment benefits and job transition assistance, can cushion the impact of job displacement. Effective social policies can help in the smoother reallocation of labor in the economy (Manyika et al., 2017).

2. **Interplay Between Economic and Policy Factors**:

- ○ **Policymaking in Response to Economic Changes**: Policymakers often react to the economic impacts of AI with measures aimed at either stimulating AI growth or protecting jobs, which in turn affects the pace and nature of job displacement (Autor et al., 2016).
- ○ **Economic Outcomes of Policy Decisions**: Conversely, policy decisions directly influence economic outcomes, such as job market dynamics and business investments in AI. For instance, stringent AI regulations might slow down automation but preserve certain job categories for longer periods (Acemoglu & Restrepo, 2018).

In summary, the interplay between economic and policy factors significantly influences the trajectory and impact of AI-induced job displacement. Understanding and strategically addressing these factors can help mitigate the adverse effects on the workforce and leverage AI's potential for economic growth and development.

Skills Gap and Re-skilling:

A significant challenge is the skills gap. Many workers displaced by AI might not have the skills needed for the new jobs created. Brynjolfsson and McAfee (2014) stress the importance of re-skilling and education programs to prepare the workforce for an AI-driven economy.

The conversation surrounding AI's impact on the job market necessitates a deep dive into the skills gap and re-skilling challenges, backed by scholarly references. This expanded analysis provides a more nuanced understanding of these issues.

1. **Understanding the Skills Gap:**
   - **Nature of the Gap**: As AI and automation technologies evolve, there's a growing discrepancy between the skills possessed by the workforce and those demanded by the new job landscape. This gap is not just about technical skills but also encompasses cognitive and socio-behavioral skills (Bughin et al., 2018).
   - **Sector-Specific Skills Demands**: Different sectors experience varied skills demands due to AI adoption. For instance, the IT sector sees a demand for advanced technical skills, while the service sector may require more socio-behavioral skills (World Economic Forum, 2020).

2. **Challenges and Strategies in Re-skilling:**
   - **Assessment of Re-skilling Needs**: Accurately assessing the re-skilling needs of the workforce is a significant challenge. This involves understanding the evolving job roles and the specific skills required for these roles (Brynjolfsson & McAfee, 2014).
   - **Effective Re-skilling Programs**: Implementing effective re-skilling programs requires collaboration between governments, educational institutions, and businesses. These

programs must be accessible, inclusive, and aligned with future job market needs (Schwab, 2016).

- ○ **Continuous Learning Culture**: Fostering a culture of continuous learning is essential to keep the workforce adaptable and agile. Lifelong learning should be integrated into the career trajectory of individuals (Bughin et al., 2018).

3. **Policy and Economic Implications**:

- ○ **Government Policy in Supporting Re-skilling**: Government policies play a critical role in supporting re-skilling initiatives through funding, incentives, and creating educational frameworks that are responsive to the changing job market (Autor et al., 2016).
- ○ **Economic Benefits of Bridging the Skills Gap**: Successfully bridging the skills gap can lead to economic benefits such as increased productivity, innovation, and a more robust job market (Manyika et al., 2017).

In conclusion, addressing the skills gap and re-skilling challenges is imperative in the era of AI-induced job displacement. Strategic collaboration among various stakeholders and a commitment to lifelong learning are key to turning these challenges into opportunities for workforce enhancement and economic growth.

Long-Term Employment Trends:

Historical trends show that technological disruptions lead to shifts in employment sectors rather than net job losses (Mokyr, Vickers, & Ziebarth, 2015). The shift towards service and knowledge-based industries post-industrial automation is an example.

The examination of long-term employment trends in the era of AI and automation provides insights into how job markets are evolving. This expanded analysis, supported by scholarly references, delves deeper into these trends.

1. **Shifting Employment Patterns:**
   - **Sectoral Shifts:** Historically, technological advancements have led to a shift from agriculture to manufacturing and then to service-based economies. Similarly, AI is anticipated to drive a shift towards more knowledge-intensive sectors (Acemoglu & Restrepo, 2020).
   - **Job Creation in Emerging Sectors:** While AI displaces certain job roles, it also creates new opportunities, especially in sectors like AI development, data analysis, and cybersecurity (World Economic Forum, 2020).
2. **The Role of AI in Job Transformation:**
   - **Enhancing Existing Roles:** AI is not just about replacing jobs; it's also about enhancing existing roles, making them more efficient and less monotonous (Brynjolfsson & McAfee, 2014).
   - **Creation of Entirely New Job Categories:** As AI evolves, entirely new job categories are emerging, which were previously unimagined, further diversifying employment opportunities (Manyika et al., 2017).
3. **The Impact of AI on Workforce Demographics:**
   - **Age and AI:** Older workers may face challenges in adapting to new technologies, necessitating targeted re-skilling programs (Autor et al., 2016).
   - **Gender and AI:** AI and automation could exacerbate existing gender employment gaps if proactive measures are not taken (Goldin, 2014).
4. **Economic and Policy Implications:**
   - **Economic Growth and Job Market Dynamics:** Long-term, AI is expected to contribute to economic growth, which could lead to a net increase in job opportunities, albeit in different sectors and roles (Bughin et al., 2018).

○ **Policy Implications**: Policies need to be adaptive and forward-thinking, focusing on education, training, and social safety nets to accommodate these shifts (Schwab, 2016).

In summary, while AI and automation are transforming the job market, they are not unilaterally eradicating jobs but rather shifting the nature and landscape of work. This transformation requires proactive adaptation in terms of policy, education, and workforce development to harness the potential benefits and mitigate the challenges posed by these technological advancements.

The narrative of AI-induced job displacement is nuanced, with realities that challenge simplistic myths. Understanding this issue requires a multi-dimensional analysis that includes technological capabilities, nature of tasks, economic incentives, policy frameworks, and the evolving nature of skills in the labor market.

Analyzing data: Job losses attributed to AI:

The analysis of job losses due to AI is a critical component in understanding the broader impact of technology on the workforce. This expansion, supported by scholarly references, provides a more in-depth look into this aspect.

Quantifying AI-Related Job Losses:

1. **Statistical Data:**

Empirical data indicates a correlation between AI integration and job losses in specific sectors. For instance, in manufacturing, automation has led to a decrease in routine manual jobs (Frey & Osborne, 2017).

The quantification of job losses attributed to AI is a complex and evolving field, requiring detailed statistical analysis to understand the full impact. This expanded discussion, supported by scholarly references, delves deeper into the statistical data surrounding AI-related job losses.

1. **Refining Statistical Models**:
   - **Incorporating AI Adoption Rates**: Advanced statistical models now factor in the rate of AI adoption across industries to predict job displacement more accurately (Autor, Levy, & Murnane, 2003).
   - **Longitudinal Data Analysis**: Studies using longitudinal data provide insights into the long-term effects of AI on job markets, observing trends over decades (Acemoglu & Restrepo, 2020).
2. **Industry-Specific Data**:
   - **Manufacturing Sector Statistics**: In the manufacturing sector, statistics show a notable decline in labor-intensive roles with the rise of AI and robotics (Bessen, 2019).
   - **Service Industry Data**: Analysis of service sector data reveals a shift from routine transactional jobs to more cognitive roles due to AI integration (Autor, 2015).
3. **Regional Variability**:
   - **Localized Data Analysis**: Studies show significant regional differences in AI-induced job losses, with some areas experiencing higher displacement rates due to their industrial makeup (Dauth et al., 2017).
   - **Global Perspective**: International data reveal varying impacts of AI on job markets in different countries, influenced by economic structures and government policies (Manyika et al., 2017).
4. **Demographic Disparities**:
   - **Impact on Different Skill Levels**: Data indicates that low-skilled workers face a higher risk of job displacement due to AI, compared to those in high-skilled roles (Chui, Manyika, & Miremadi, 2016).

- **Age and Gender Factors**: Statistical trends also show variations in AI-related job displacement based on age and gender, with certain demographics being more vulnerable (Korinek & Stiglitz, 2017).

In summary, statistical data plays a crucial role in understanding AI-related job losses. It reveals the nuances of AI's impact across different industries, regions, and demographic groups. This refined understanding is essential for policymakers and stakeholders to address the challenges posed by AI in the labor market effectively.

II. Projected Trends:

Projections suggest that up to 30% of jobs could be at high risk of automation by the mid-2030s, with variations depending on the sector and geographical location (PwC, 2018).

The projection of AI-related job losses involves complex forecasting models that take into account various factors such as technological advancements, industry adaptation rates, and policy interventions. This expanded discussion, supported by scholarly references, explores the projected trends in AI-related job losses.

1. **Forecasting Models**:
    - **Econometric Predictions**: Advanced econometric models are used to predict the future impact of AI on jobs, incorporating variables like technology adoption rates, industry growth, and labor market dynamics (Frey & Osborne, 2017).
    - **Scenario-Based Projections**: Researchers use scenario-based analyses to envision various futures, ranging from minimal impact to extensive job losses, depending on the speed and nature of AI integration (Arntz, Gregory, & Zierahn, 2016).
2. **Sector-Specific Projections**:

- **Automated Manufacturing**: Projections suggest a significant reduction in manual jobs in manufacturing, but an increase in technical and supervisory roles (Manyika et al., 2017).
- **Service and Retail**: In service sectors, AI is expected to automate routine tasks, but also create new opportunities in customer service and personalization (Bessen, 2019).

3. **Regional and Global Trends**:

- **Developed vs. Developing Economies**: Projections indicate that developed economies might experience faster job displacement due to higher AI adoption, whereas in developing countries, the impact may be gradual (World Economic Forum, 2018).
- **International Competitiveness**: Countries investing heavily in AI are projected to gain a competitive edge, potentially leading to job creation in AI-driven industries (Acemoglu & Restrepo, 2020).

4. **Impact on Workforce Demographics**:

- **Skilled vs. Unskilled Labor**: Forecasts suggest that low-skilled jobs are at higher risk of displacement, while demand for high-skilled workers, especially in AI and data analysis, may increase (Chui, Manyika, & Miremadi, 2016).
- **Age and Gender Implications**: Different age groups and genders may be affected disproportionately, with younger workers potentially adapting more quickly to AI-driven changes (Korinek & Stiglitz, 2017).

In summary, projected trends in AI-related job losses are a blend of econometric modeling, scenario analysis, and sector-specific forecasts. These projections are vital for understanding the potential paths AI development might take and the corresponding implications for the global workforce. Policymakers, educators, and industry leaders can use this information to prepare for and mitigate the potential negative impacts of AI on employment.

Sector-Specific Impact:

1. Manufacturing and Automation:

The manufacturing sector has seen significant job losses due to automation, primarily in routine, repetitive tasks (Autor, 2015).

The impact of AI and automation on job losses in the manufacturing sector is a significant area of study. This expanded analysis examines the nuances of this impact, supported by appropriate academic references.

1. **Automation Trends in Manufacturing**:
   - **Robotics and Automation**: The integration of robotics and AI in manufacturing has led to increased efficiency but also a reduction in certain types of jobs, particularly in assembly and production lines (Manyika et al., 2017).
   - **Process Optimization**: AI-driven process optimization has resulted in the need for fewer workers in roles that involve repetitive tasks, but has also increased demand for skilled technicians and engineers (Ford, 2015).
2. **Impact on Employment**:
   - **Job Displacement**: Studies indicate that automation in manufacturing predominantly affects low-skill and routine jobs, leading to significant displacement in these areas (Acemoglu & Restrepo, 2020).
   - **Job Creation**: Conversely, there is evidence of job creation in areas related to AI and automation, such as maintenance, programming, and system management (Bessen, 2019).
3. **Economic and Geographic Considerations**:

- **Economic Efficiency vs. Employment**: While automation contributes to economic efficiency and productivity, it presents challenges in terms of employment, especially in regions heavily dependent on manufacturing jobs (Autor, 2015).
- **Global Distribution of Manufacturing**: The automation trend might lead to a reshaping of the global manufacturing landscape, with some jobs returning to developed countries due to the reduced cost advantage of labor in developing countries (Baldwin, 2019).

4. **Future Outlook**:

- **Advanced Manufacturing Technologies**: Emerging technologies like 3D printing and advanced robotics are expected to further transform manufacturing, potentially leading to more complex job displacement patterns (Schwab, 2017).
- **Reskilling and Workforce Adaptation**: The future of manufacturing employment will likely require significant investments in reskilling and education to prepare workers for the changing nature of manufacturing jobs (WEF, 2018).

II. Service Sector Vulnerability:

Roles in customer service and data entry in the service sector are also increasingly susceptible to AI-driven automation (Manyika et al., 2017).

The service sector's vulnerability to AI-induced job losses necessitates a detailed analysis. This expansion focuses on various dimensions of how AI impacts jobs in the service sector, supported by academic references.

1. **AI and Automation in Service Jobs**:

- **Customer Service Automation**: AI technologies, particularly chatbots and automated response systems, have increasingly replaced human roles in customer service (Bughin et al., 2018).
- **Retail and E-Commerce**: AI-driven systems in e-commerce, like recommendation engines, have reduced the need for sales staff in traditional retail settings (Huang & Rust, 2018).

2. **Impact on Employment Patterns**:
   - **Job Displacement**: There is evidence of significant job displacement in roles that are repetitive and predictable, such as data entry and basic customer service (Frey & Osborne, 2017).
   - **Job Transformation**: Some service sector jobs are transforming, requiring employees to develop new skills to work alongside AI technologies (Autor, 2015).

3. **Economic Implications**:
   - **Productivity vs. Employment**: While AI enhances productivity and efficiency in the service sector, it poses challenges for employment, particularly for low-skill labor (Susskind & Susskind, 2015).
   - **Shift in Skill Demand**: The demand for social, emotional, and higher cognitive skills is rising, as these are less susceptible to automation (Brynjolfsson & McAfee, 2014).

4. **Sector-Specific Examples**:
   - **Financial Services**: AI in banking and finance has automated tasks like transaction processing, but also created new roles in AI oversight and algorithm management (Arntz, Gregory, & Zierahn, 2016).
   - **Healthcare Services**: In healthcare, AI has automated some diagnostic procedures, but it has also expanded roles for medical professionals in interpreting AI outputs and providing personalized care (Jiang et al., 2017).

5. **Future Outlook and Policy Recommendations**:
    ◦ **Adaptive Workforce**: Continuous learning and adaptability are key for workers in the service sector to remain relevant (WEF, 2018).
    ◦ **Policy Interventions**: Governments and organizations must invest in education and training programs to prepare the workforce for an AI-driven service sector (Manyika et al., 2017).

Discrepancy in Impact Across Demographics:

1. **Impact on Low-Skill Jobs**:

Lower-skilled jobs are more vulnerable to automation, leading to higher job displacement rates in this demographic (Acemoglu & Autor, 2011).

The differential impact of AI on various demographic groups, particularly concerning low-skill jobs, is a critical aspect of the AI-induced job displacement conversation. This expansion delves deeper into how AI disproportionately affects low-skill job sectors, supported by academic references.

1. **Increased Vulnerability of Low-Skill Jobs**:
    ◦ **High Automation Potential**: Low-skill jobs, characterized by repetitive and routine tasks, have a higher automation potential, leading to a greater risk of displacement (Frey & Osborne, 2017).
    ◦ **Limited Job Mobility**: Individuals in low-skill jobs often face barriers to job mobility due to a lack of requisite skills for transitioning into AI-resistant roles (Autor, 2015).
2. **Socioeconomic Implications**:
    ◦ **Widening Income Inequality**: The automation of low-skill jobs can exacerbate income inequality, as displaced

workers may struggle to find equivalent roles (Atkinson & Wu, 2017).

- ○ **Economic Insecurity**: Job displacement in low-skill sectors contributes to economic insecurity and can have cascading effects on health and social stability (Case & Deaton, 2020).

3. **Demographic Disparities**:

- ○ **Disproportionate Impact on Certain Groups**: Certain demographic groups, such as those with lower educational attainment or from economically disadvantaged backgrounds, are more likely to be employed in low-skill jobs and thus more vulnerable to AI-driven displacement (Acemoglu & Restrepo, 2020).
- ○ **Gender and Race Disparities**: There are also disparities in the impact of AI on low-skill jobs based on gender and race, with some studies indicating women and minority groups may be disproportionately affected (Buolamwini & Gebru, 2018).

4. **Policy and Educational Interventions**:

- ○ **Reskilling Initiatives**: Governments and organizations need to prioritize reskilling and upskilling initiatives to enable vulnerable workers to transition to new roles (Schwab, 2017).
- ○ **Inclusive AI Development**: Policies should focus on inclusive AI development that considers the needs and abilities of all demographic groups, thereby reducing the risk of exacerbating existing inequalities (WEF, 2018).

5. **Future Outlook**:

- ○ **Shift in Labor Market Dynamics**: The labor market is expected to shift, with an increase in demand for jobs requiring social, emotional, and higher cognitive skills (Bughin et al., 2018).

- **Long-Term Societal Impact**: The long-term societal impact of AI on low-skill jobs underscores the need for proactive measures to mitigate risks and harness opportunities (Manyika et al., 2017).

In conclusion, the impact of AI on low-skill jobs is significant and multifaceted, with pronounced disparities across demographic groups. This necessitates targeted policy and educational interventions to mitigate adverse effects and promote an inclusive transition in the evolving job market.

II. Geographical Variations:

Job losses due to AI vary significantly across different regions, with certain areas being more susceptible due to their economic structure (Autor et al., 2016).

The impact of AI on job displacement exhibits significant geographical variations, which can be analyzed further in terms of regional economic structures, technological adoption rates, and policy frameworks. This expansion provides a deeper insight into these variations, supported by academic sources.

1. **Regional Economic Structures**:
   - **Diverse Industrial Profiles**: Different regions have varying industrial profiles, with some areas more focused on manufacturing and others on services or technology, influencing the impact of AI on job markets (Autor, Levy, & Murnane, 2003).
   - **Resource Distribution**: Regions rich in natural resources may experience less immediate impact from AI in certain sectors, compared to urban areas with a high concentration of service and tech industries (Moretti, 2012).
2. **Technological Adoption Rates**:
   - **Infrastructure and Investment**: Regions with better technological infrastructure and higher investment in AI

and automation technologies are likely to experience more rapid job displacement (Brynjolfsson & McAfee, 2014).

- **Rural vs. Urban Divide**: Urban areas often have higher rates of AI adoption due to better infrastructure and higher demand for efficiency, contrasting with rural areas where technology penetration is slower (Partridge, 2015).

3. **Policy Frameworks**:

- **Government Initiatives and Support**: Regions with proactive government policies supporting technological advancement and workforce transition are better equipped to handle AI-induced job displacement (Acemoglu & Restrepo, 2018).

- **Educational and Training Programs**: The availability and quality of educational and training programs for workforce reskilling also vary geographically, impacting the ability of the workforce to adapt to AI-induced changes (Goldin & Katz, 2008).

4. **Global North vs. Global South Disparities**:

- **Economic Development Level**: Countries in the Global North, with advanced economies, are adopting AI at a faster pace compared to the Global South, leading to differing impacts on job displacement (Lee, 2018).

- **Dependency on Certain Sectors**: Some regions, especially in the Global South, have economies heavily reliant on sectors like agriculture, which are less immediately impacted by AI compared to manufacturing and services predominant in the Global North (Rodrik, 2016).

5. **Long-term Regional Shifts**:

- **Changing Economic Landscapes**: Long-term, regions may witness shifts in their economic landscapes, with some areas becoming hubs for AI development and others facing economic challenges due to job losses in traditional sectors (Katz & Margo, 2014).

In summary, geographical variations in the impact of AI on job displacement are influenced by multiple factors, including regional economic structures, technological adoption rates, policy frameworks, and educational and training opportunities. These variations necessitate region-specific strategies to address the challenges and opportunities presented by AI-induced job market changes.

The Role of Economic Factors:

1. **Economic Cycles and AI Adoption**:

Economic downturns often accelerate the adoption of AI and automation as businesses seek cost savings, potentially leading to increased job losses (Brynjolfsson & McAfee, 2014).

The relationship between economic cycles and AI adoption, particularly in the context of job displacement, is complex and multifaceted. This expansion offers a deeper understanding, supported by academic references.

1. **AI Adoption During Economic Booms**:
   - **Increased Investment**: During economic upturns, businesses are more likely to invest in AI and automation technologies, aiming to enhance productivity and competitive advantage (Bughin, Hazan, Ramaswamy, Chui, Allas, Dahlström, Henke, & Trench, 2018).
   - **Amplified Job Displacement**: This increased investment can lead to amplified displacement, particularly in sectors where AI can easily replace routine tasks (Autor, Dorn, & Hanson, 2015).
2. **AI Adoption in Economic Downturns**:
   - **Cost-Cutting Measures**: In recessions, companies often turn to AI as a cost-cutting measure, replacing more expensive human labor to maintain profitability (Kaplan & Haenlein, 2019).

- **Cyclical vs. Structural Unemployment**: Distinguishing between cyclical unemployment, due to economic downturns, and structural unemployment, due to AI adoption, is crucial in policy formulation (Acemoglu & Restrepo, 2020).

3. **Long-Term Economic Impact**:
    - **Productivity Gains**: AI has the potential to drive long-term productivity gains, which can lead to economic growth and job creation in new sectors (Brynjolfsson, Rock, & Syverson, 2018).
    - **Inequality Concerns**: However, these gains may be unevenly distributed, exacerbating income and wealth inequalities (Piketty, 2014).

4. **Sector-Specific Dynamics**:
    - **Varying Impact Across Industries**: The impact of AI on job displacement varies significantly across different industries, depending on the nature of work and the feasibility of automation (Manyika et al., 2017).

5. **Policy and Economic Responses**:
    - **Government Interventions**: Effective government policies are essential to mitigate the negative impacts of AI on employment, especially during economic downturns (Autor & Dorn, 2013).
    - **Reskilling and Safety Nets**: Investing in education, training, and social safety nets is crucial to support workers displaced by AI (Schwab, 2016).

In summary, the role of economic cycles in AI adoption and the consequent impact on job displacement is influenced by investment patterns, cost-cutting strategies, and policy responses. Understanding these dynamics is critical for developing strategies that not only leverage AI for economic growth but also address the challenges it poses to the workforce.

II. Competitive Pressures:

In highly competitive industries, the pressure to adopt AI and reduce labor costs can contribute to job losses (Bughin et al., 2018).

The influence of competitive pressures on AI adoption and its consequent impact on job displacement is a significant aspect of the AI impact narrative. This expanded analysis delves deeper into this aspect, supported by academic references.

1. **Competitive Pressure as a Driver for AI Adoption:**
   - Businesses increasingly adopt AI to gain competitive advantages, such as cost reduction, efficiency improvements, and innovation in products and services (Porter & Heppelmann, 2014).
   - This pressure can lead to a 'race to the bottom' in labor standards, as companies strive to cut costs, potentially leading to job displacement (Srnicek, 2017).
2. **Impact on Small and Medium Enterprises (SMEs):**
   - SMEs often face difficulties in adopting AI due to limited resources, potentially leading to a competitive disadvantage and job losses in these firms (Audretsch, 2015).
   - Conversely, AI can provide SMEs with tools to compete against larger firms, potentially preserving jobs by enhancing productivity and market reach (Farrell & Greig, 2016).
3. **Global Competitive Dynamics:**
   - The global nature of competition means that companies in developed countries may adopt AI more rapidly to compete with low-wage labor in developing countries, influencing job markets worldwide (Baldwin, 2019).
4. **Industry-Specific Competitive Forces:**
   - In industries where AI can significantly enhance productivity or innovation, such as manufacturing or tech, competitive pressures can lead to more aggressive AI adoption and potential job displacement (Ford, 2015).

5. **Long-Term Competitive Strategy**:
   - Companies adopting AI for short-term gains may overlook long-term implications on employment and skills development, potentially harming their competitiveness in the future (Brynjolfsson & McAfee, 2014).

6. **Policy and Regulation**:
   - Governments and regulatory bodies play a crucial role in shaping how competitive pressures impact job displacement through AI, by setting standards and providing incentives for responsible AI adoption (Kenney & Zysman, 2016).

In conclusion, competitive pressures play a vital role in shaping AI adoption and its impact on job displacement. This complex interplay of factors includes the need for short-term competitive advantages, the challenges faced by SMEs, the global nature of competition, industry-specific dynamics, long-term strategic considerations, and the role of policy and regulation in mitigating adverse effects on the job market. Understanding these factors is crucial for businesses, policymakers, and workers navigating the evolving landscape of AI and employment.

Roles AI cannot replace: The human touch:

In discussing roles that AI cannot replace, it is crucial to emphasize the unique aspects of human interaction, empathy, and creativity that AI systems currently lack. This expanded analysis, supported by academic references, further explores the importance of the human touch in various roles.

1. Healthcare and Patient Interaction:

The importance of the human touch in healthcare and patient interaction is a critical aspect where AI cannot fully replicate human capabilities. This expanded analysis focuses on specific areas within healthcare where the human touch is irreplaceable, supported by academic references.

1. **Empathy in Patient Care**:
   - Empathy plays a fundamental role in patient care, significantly impacting patient satisfaction, adherence to treatment, and overall health outcomes. The ability of healthcare professionals to show empathy, understand patient concerns, and provide emotional support is something AI cannot mimic (Halpern, 2001).
   - Emotional intelligence in healthcare providers facilitates better patient-provider relationships, which is vital for effective treatment and care (Mayer & Salovey, 1997).
2. **Complex Clinical Judgments**:
   - Healthcare professionals often make complex clinical judgments that involve not only technical knowledge but also ethical considerations, cultural sensitivity, and patient preferences (Pellegrino & Thomasma, 1993).
   - While AI can support data-driven decisions, it lacks the ability to navigate the moral and ethical dimensions of clinical care (Char, 2018).
3. **Communication Skills**:
   - Effective communication skills are essential in healthcare, particularly in delivering bad news or discussing sensitive health issues. The nuanced understanding and compassionate delivery of such information require a human touch (Fallowfield & Jenkins, 2004).
   - AI lacks the ability to provide the warmth, reassurance, and personal connection that come with human interaction (Back et al., 2009).
4. **Holistic Care**:

- ○ Holistic care involves considering the physical, emotional, social, and spiritual needs of patients. Healthcare professionals' ability to provide holistic care is grounded in human empathy and understanding (Dossey & Keegan, 2013).
- ○ AI tools may assist in certain aspects of care, but they cannot fully understand and integrate these diverse human needs (McSherry & Ross, 2010).

While AI can augment certain aspects of healthcare, it cannot replace the human touch in areas such as empathy, complex clinical judgments, effective communication, and holistic care. The nuanced, empathetic, and ethically complex nature of these interactions underscores the irreplaceable value of human healthcare providers in patient care and interaction

II. Education and Learning:

The role of the human touch in education and learning is a significant area where AI cannot fully take over. This expanded analysis delves into specific aspects of education where human interaction is crucial, supported by academic references.

1. **Emotional Support and Motivation**:
   - ○ Teachers play a critical role in providing emotional support and motivation to students. This includes recognizing and responding to students' emotional states, something AI is not equipped to do (Hattie, 2009).
   - ○ The motivational impact of a teacher can significantly influence student engagement and academic performance (Ryan & Deci, 2000).
2. **Adaptation to Individual Learning Styles**:
   - ○ Educators are adept at adapting teaching strategies to cater to the diverse learning styles and needs of students. AI

lacks the nuanced understanding required to tailor learning experiences at this individual level (Gardner, 1983).

○ Personalized attention helps in addressing specific learning challenges, which is beyond the capability of current AI technologies (Tomlinson & McTighe, 2006).

3. **Socio-Emotional Learning (SEL)**:

○ SEL is crucial for students' emotional and social development. Teachers facilitate SEL through personal interactions, classroom discussions, and group activities, which AI cannot replicate (Durlak et al., 2011).

○ The role of teachers in nurturing social skills, empathy, and emotional intelligence is indispensable (Elias et al., 1997).

4. **Mentorship and Role Modeling**:

○ Teachers often serve as mentors and role models, influencing students' attitudes, values, and future aspirations. The inspirational role of a teacher in shaping a student's life path cannot be replaced by AI (Bandura, 1977).

○ Positive teacher-student relationships are linked to better academic outcomes and personal development (Roorda et al., 2011).

5. **Facilitating Critical Thinking and Creativity**:

○ Human educators play a key role in fostering critical thinking and creativity. They challenge students' thinking, stimulate discussions, and encourage creative problem-solving (Paul & Elder, 2006).

○ AI lacks the ability to effectively promote these higher-order thinking skills that are essential for students' overall intellectual development (Sawyer, 2006).

While AI can provide valuable tools in education, it cannot replace the human touch in providing emotional support, adapting to individual learning styles, fostering socio-emotional learning, acting as mentors and role models, and facilitating critical thinking and creativity. The unique and irreplaceable role of educators in these aspects highlights the importance of human interaction in the educational process.

III. Creative Industries:

The role of human creativity in the creative industries is a critical area where AI cannot fully substitute. This expanded analysis focuses on the unique aspects of human creativity in these industries, supported by academic references.

1. **Originality and Emotional Depth:**
   - Human creativity is characterized by originality and emotional depth, which AI cannot replicate. The creative process involves not just the generation of new ideas but also the emotional and cultural significance behind them (Csikszentmihalyi, 1996).
   - Human artists convey complex emotions and experiences through their work, something that AI, which lacks personal experiences and emotions, cannot achieve (Gardner, 1982).

2. **Cultural Context and Relevance:**
   - Creativity is deeply embedded in cultural context. Artists draw upon their cultural backgrounds and societal experiences to create works that resonate with human audiences (Bourdieu, 1993).
   - AI lacks the ability to understand and integrate cultural nuances and historical context in the way humans do (Hall, 1976).

3. **Innovation and Breaking Norms:**
   - Human creativity often involves breaking norms and introducing innovative concepts. Artists and creators frequently challenge existing paradigms and push boundaries, a quality not inherent in AI (Kuhn, 1962).

- AI tends to generate outputs based on existing data and patterns, which may limit its capacity for true innovation (Boden, 2004).

4. **Interpersonal Collaboration and Exchange**:

- Creative industries often thrive on collaboration and interpersonal exchange. The synergy between different human creators leads to the emergence of new ideas and styles (John-Steiner, 2000).
- While AI can assist in the creative process, it cannot replicate the dynamic and organic nature of human collaboration (Sawyer, 2007).

5. **Audience Engagement and Interpretation**:

- Art and creative works often invite subjective interpretation and personal engagement from the audience. Human creators design experiences that resonate on a personal level with viewers or listeners (Dewey, 1934).
- AI lacks the ability to create works that engage audiences in a deeply personal and subjective manner (Gombrich, 1960).

While AI can support various tasks in the creative industries, it cannot replace the human touch in aspects like originality, cultural relevance, innovation, collaboration, and audience engagement. The unique and irreplaceable role of human creativity in these areas underscores the importance of human artists and creators in the cultural and creative landscape.

IV. Social Work and Counseling:

The indispensability of human intervention in social work and counseling, even in an era of advanced AI, is a vital area to consider. This expanded analysis delves into the unique aspects of human interaction in these fields, supported by relevant academic references.

1. **Empathy and Emotional Intelligence**:

- Empathy and emotional intelligence are core components of effective social work and counseling. These human qualities enable professionals to understand and resonate with clients' emotions and situations (Rogers, 1961).
- AI, despite its advancements, lacks the genuine empathy and emotional intelligence necessary to fully comprehend and respond to the nuanced emotional states of individuals (Salovey & Mayer, 1990).

2. **Building Trust and Therapeutic Relationships:**

- The establishment of trust and a therapeutic relationship between the counselor and the client is fundamental to successful outcomes. Human counselors are adept at building these relationships through interpersonal skills and genuine care (Yalom, 2002).
- AI systems may assist in some aspects of therapy but lack the ability to create deep, trust-based relationships essential for effective therapy (Frank & Frank, 1991).

3. **Cultural Sensitivity and Contextual Understanding:**

- Social workers and counselors often work with diverse populations, requiring a deep understanding of cultural and contextual factors. This understanding helps in tailoring interventions to be culturally sensitive and relevant (Sue & Sue, 2012).
- AI cannot fully grasp the complex cultural nuances and contextual subtleties that human professionals can, which are essential in counseling and social work (Pedersen, 1991).

4. **Dealing with Complex Human Situations:**

- Social work and counseling often involve dealing with complex, unpredictable human situations. Humans have the ability to adapt and respond to such complexities in a flexible and understanding manner (Skovholt & Trotter-Mathison, 2016).

- AI, guided by algorithms and data, may not effectively handle the unpredictable and deeply personal aspects of human issues (Norcross & Wampold, 2011).

5. **Ethical Decision-Making**:

- Ethical decision-making in social work and counseling involves nuanced judgment calls, often in ethically ambiguous situations. Human professionals can navigate these complexities with moral and ethical consideration (Corey, Corey, & Callanan, 2014).
- AI systems, while they can follow ethical guidelines, lack the capacity for moral reasoning and the nuanced judgment required in these fields (Beauchamp & Childress, 2012).

While AI can offer certain tools and support in the fields of social work and counseling, it cannot replace the human qualities of empathy, trust-building, cultural sensitivity, adaptability, and ethical judgment. These human attributes are essential for effectively addressing the complex and nuanced needs of clients in these fields.

V. Leadership and Management:

The role of human touch in leadership and management remains crucial, even in an era increasingly influenced by artificial intelligence. This expanded analysis explores why human qualities are irreplaceable in these areas, with support from academic references.

1. **Emotional Intelligence and Leadership**:

- Emotional intelligence is a key component of effective leadership. It involves understanding and managing one's own emotions and empathizing with others, a trait AI cannot replicate (Goleman, 1995).
- Leaders with high emotional intelligence are better equipped to motivate, inspire, and connect with their

teams, fostering a positive work environment (Mayer, Roberts, & Barsade, 2008).

2. **Complex Decision-Making**:
   - Leadership often requires making complex decisions that involve ethical considerations, long-term strategic thinking, and understanding the human impact of these decisions (Bass & Bass, 2008).
   - AI, though capable of processing vast amounts of data, lacks the ability to make nuanced decisions that consider both logical and emotional aspects (Kahneman, 2011).

3. **Innovation and Creativity**:
   - Innovation and creativity are essential in leadership and management. Humans excel at thinking outside the box, generating novel ideas, and inspiring creativity in others (Amabile, 1996).
   - AI can support innovation through data analysis and pattern recognition, but it cannot replicate the human capacity for creative thinking and ideation (Florida, 2012).

4. **Building and Maintaining Relationships**:
   - Building and maintaining relationships is central to effective management. This includes networking, mentorship, and fostering a culture of trust and respect (Bennis, 2009).
   - While AI can assist in communication processes, it cannot replace the personal touch and genuine connections formed in human relationships (Uzzi, 1997).

5. **Adaptability and Flexibility**:
   - Effective leaders must be adaptable and flexible, responding to changing environments and unexpected challenges (Yukl, 2009).
   - AI systems follow predefined algorithms and lack the ability to adapt spontaneously to unforeseen circumstances in the way humans can (Weick & Quinn, 1999).

6. **Ethical Leadership**:

- Ethical leadership is vital for maintaining integrity and moral standards within organizations. This involves making decisions that are not only legally compliant but also ethically sound (Brown & Treviño, 2006).
- AI, though capable of following ethical guidelines, cannot engage in moral reasoning or understand the complex ethical dilemmas often faced by leaders (Kidder, 2009).

Finally, leadership and management roles significantly benefit from the human touch. Emotional intelligence, complex decision-making, innovation, relationship-building, adaptability, and ethical considerations are aspects where human leaders excel and AI cannot adequately substitute. These human attributes are integral to the essence of effective leadership and management.

In conclusion, while AI technologies offer remarkable capabilities in various fields, they cannot replace the human touch in roles that require empathy, creativity, emotional intelligence, and subjective judgment. From healthcare and education to creative industries and leadership, the intrinsic human qualities that define these roles remain irreplaceable. Understanding the limitations of AI in these contexts is vital for maintaining the balance between technological advancement and the preservation of uniquely human skills and interactions.

The cycle of technology: Job destruction and creation:

The relationship between technological advancement and job displacement is a cyclical one, characterized by both job destruction and job creation. This expanded analysis delves deeper into this cycle, supported by academic references.

1. **Historical Perspective**:

Understanding the historical perspective of the relationship between technology and job displacement is crucial for contextualizing contemporary concerns. This expanded analysis delves deeper into this historical aspect, using relevant academic sources.

1. **The Industrial Revolution**:
   - The Industrial Revolution is a pivotal example of technological disruption leading to job displacement and creation. Craftspeople and artisans saw their roles diminished with the advent of machines, but new roles in factories and urban centers emerged (Mokyr, 2002).
   - Landes (2003) notes that while the Industrial Revolution led to short-term job losses, it ultimately boosted productivity, wages, and living standards, laying the foundation for modern economies.

2. **The Transition from Agriculture to Industry**:
   - The shift from agrarian societies to industrial ones saw a significant migration of labor. Boserup (1965) discusses how technological advances in agriculture reduced the need for labor, pushing people towards industrial jobs.
   - Federico (2005) highlights the role of technological advancements in agriculture in facilitating this transition, thereby reshaping entire labor markets and societal structures.

3. **The Introduction of Computers and Automation**:
   - The latter half of the 20th century saw computers and automation begin to transform workplaces. Autor, Levy, and Murnane (2003) demonstrate how this led to an increase in demand for skilled labor while reducing demand for routine manual tasks.
   - Goldin and Katz (2008) analyze how the introduction of computers altered the skill composition of the workforce, emphasizing the need for education and skill development.

4. **Technological Complementarity and Substitution**:
   - Bresnahan and Trajtenberg (1995) introduce the concept of "general purpose technologies" (GPTs), like the steam engine or computer, which are transformative and have broad-ranging effects on various sectors, leading to both job displacement and creation.
   - David (1990) discusses how each major technological advancement, while initially disruptive, eventually becomes complementary to human labor, enhancing productivity and creating new employment opportunities.

5. **Long-term Economic Impacts**:
   - A long-view analysis by Crafts (2004) suggests that while technology displaces specific jobs in the short term, it tends to lead to higher overall employment and better living standards in the long term.
   - Atkinson and Wu (2017) argue that technological progress, despite its disruptive effects, has been a key driver of economic growth and job creation over the centuries.

II. Job Destruction:

The phenomenon of job destruction as a consequence of technological advancement is a critical aspect of the broader narrative on technology's impact on employment. This expanded analysis examines the nuances and complexities of job destruction, supported by scholarly sources.

1. **Nature of Job Destruction**:
   - Acemoglu and Restrepo (2018) argue that technological innovations often lead to a decline in certain types of jobs, particularly those involving routine tasks that can be easily automated.
   - Brynjolfsson and McAfee (2014) highlight how digital technologies, unlike previous technological advancements,

have the potential to replace a wide range of jobs, from manual labor to cognitive tasks.

2. **Displacement of Skilled Labor**:
   - Autor (2015) notes that technological advancements not only impact low-skill jobs but increasingly threaten middle-skill jobs, leading to job polarization.
   - Goos and Manning (2007) discuss the 'hollowing out' of the job market, where middle-wage, middle-skill jobs are the most susceptible to automation.

3. **Sector-Specific Impacts**:
   - Manyika et al. (2017) provide an analysis of how automation impacts various sectors differently, with manufacturing, retail, and transportation seeing significant job losses due to automation technologies.
   - Bessen (2016) examines the historical and contemporary impacts of automation in manufacturing, indicating a trend of declining employment in this sector due to technological advancements.

4. **Geographical and Demographic Variations**:
   - Autor and Dorn (2013) highlight how job destruction due to technology varies significantly across regions, disproportionately affecting areas with industries susceptible to automation.
   - Forsythe et al. (2020) analyze the impact of automation on employment in rural vs. urban areas, showing a higher vulnerability in rural regions where routine jobs are more prevalent.

5. **Short-Term vs. Long-Term Effects**:
   - Spencer (2018) discusses the distinction between short-term job losses and long-term labor market adjustments, noting that while some jobs are permanently lost, new opportunities eventually emerge.

- Frey and Osborne (2017) argue that while technology induces job destruction in the short term, its long-term effects on job creation and economic growth can be positive.

III. Job Creation:

The dynamic of job creation amidst the rise of technology is a multifaceted aspect, integral to understanding the overall impact of technological advancements on the labor market. This expanded analysis delves into the intricacies of job creation, supported by scholarly references.

1. **New Categories of Jobs:**
   - Bughin et al. (2018) emphasize that technological advancements, especially in AI and robotics, are creating new categories of jobs, which require different skill sets compared to traditional roles.
   - World Economic Forum (2020) reports that the emergence of digital economies is fostering new roles in areas like cybersecurity, data analysis, and AI development.

2. **Enhancement of Existing Jobs:**
   - Autor (2015) discusses how technology complements certain skill sets, leading to job enhancement rather than replacement, particularly in roles that require problem-solving and creativity.
   - Agrawal, Gans, and Goldfarb (2019) highlight the augmentation of human capabilities in existing jobs through AI, enhancing efficiency and productivity.

3. **Shift in Skill Demand:**
   - The McKinsey Global Institute (2017) report indicates a shift towards higher cognitive skills, social and emotional skills, and technological skills in the workforce due to technological advancements.

- ○ Deming (2017) examines the increasing importance of soft skills in the labor market, as automation takes over more routine tasks.

4. **Geographic and Sectoral Variations**:
   - ○ Katz and Margo (2014) analyze how technological changes lead to job creation in certain geographic areas, often those with a strong base in technology and innovation.
   - ○ Atkinson and Wu (2017) discuss sectoral shifts in job creation, with sectors like healthcare, education, and professional services experiencing growth due to technological advancements.

5. **Long-Term Economic Growth**:
   - ○ Aghion et al. (2017) argue that technological progress is a key driver of long-term economic growth, which eventually leads to the creation of more jobs and higher living standards.
   - ○ Acemoglu and Restrepo (2020) contend that while there is initial displacement, the long-term equilibrium usually sees a net increase in employment due to technological advancements.

IV. Job Transformation:

The concept of job transformation, as an integral component of the cycle of technology, offers a nuanced understanding of how technological advancements reshape the nature and structure of work. This expanded analysis examines the various facets of job transformation, substantiated by scholarly references.

1. **Nature of Job Transformation**:
   - ○ Brynjolfsson and McAfee (2014) discuss how technology not only automates tasks but also transforms them, creating jobs that require new skill sets and ways of working.

- Autor (2015) points out that automation often complements rather than substitutes human labor, leading to a transformation in job roles that emphasize human-centric skills.

2. **Shift from Routine to Non-Routine Tasks:**
   - Goos, Manning, and Salomons (2014) emphasize the shift from routine to non-routine tasks in the labor market, driven by the inability of AI and automation to replicate complex human interactions and problem-solving abilities.
   - Bessen (2019) explores the evolution of job tasks from routine manual and cognitive tasks to those requiring more interpersonal and analytical skills.

3. **Role of Digital Literacy and Tech Skills:**
   - The OECD (2019) report on Skills Outlook highlights the increasing importance of digital literacy and technical skills in the transformed job landscape.
   - Van Laar et al. (2017) argue that digital skills are becoming fundamental to most jobs, necessitating continuous learning and adaptation by the workforce.

4. **Impact on Job Design and Work Structure:**
   - Valenduc and Vendramin (2017) explore how technology influences job design, leading to more flexible, but sometimes more fragmented and precarious work arrangements.
   - Manyika et al. (2017) from the McKinsey Global Institute highlight the changing nature of work, including the rise of remote work, gig economy, and project-based roles due to technological advancements.

5. **Adaptation and Skill Development:**
   - Schwab (2016) in "The Fourth Industrial Revolution" discusses the necessity for workers and organizations to adapt

to rapidly changing job requirements through upskilling and reskilling.

- Bughin et al. (2018) reinforce the need for lifelong learning and adaptability in the workforce to meet the demands of transformed job roles.

V. **Economic and Social Implications**:

The interplay of job destruction and creation due to technological advancements carries profound economic and social implications. This expanded analysis delves into these aspects, substantiated by relevant academic sources.

1. **Economic Implications**:
   - Acemoglu and Restrepo (2020) discuss the economic disruptions caused by new technologies, which can lead to short-term job losses and wage stagnation, but also long-term economic growth through new industries.
   - Mokyr, Vickers, and Ziebarth (2015) highlight that while technology-driven job destruction may cause temporary economic downturns, it eventually leads to higher productivity and economic growth.
2. **Income Inequality and Wealth Distribution**:
   - Piketty (2014) addresses the concern of increasing income inequality exacerbated by technology, as high-skill workers benefit more from technological advancements than low-skill workers.
   - Atkinson and Bourguignon (2014) explore the implications of technological change on wealth distribution, suggesting the need for policy interventions to mitigate inequality.
3. **Social Implications and Labor Market Dynamics**:
   - Autor (2019) examines the impact of automation on labor market polarization, where middle-skill jobs decline while

low-skill and high-skill jobs grow, affecting social structures and class dynamics.

- Katz and Margo (2014) study historical labor market shifts, noting that technological changes often lead to social upheaval and a redefinition of work ethics and values.

4. **Impact on Workforce Demographics**:

- Goldin and Katz (2008) analyze the transformation of workforce demographics, where the demand for more educated and skilled workers rises, impacting gender and age dynamics in the labor market.
- Chetty, Hendren, Jones, and Porter (2020) focus on the intergenerational impact of these changes, emphasizing the challenges for younger generations to adapt to a rapidly evolving job market.

5. **Policy and Educational Response**:

- Heckman and Mosso (2014) stress the importance of early childhood education and vocational training in preparing the workforce for a technologically advanced economy.
- Bresnahan and Trajtenberg (1995) argue for proactive policy measures, including job retraining and social safety nets, to support those displaced by technological changes.

VI. Policy Response:

The policy response to the challenges posed by the cycle of technology, specifically job destruction and creation, is a critical area of focus. This expanded analysis, supported by relevant academic sources, delves into the multifaceted policy approaches required to address these issues.

1. **Strategies for Workforce Reskilling and Upskilling**:

- Bessen (2019) emphasizes the importance of reskilling and upskilling programs to help workers adapt to new technologies, advocating for government and industry collaboration in providing training opportunities.

- Autor and Dorn (2013) suggest that policies should focus on lifelong learning and vocational training, ensuring that workers can continuously adapt to changing job requirements.

2. **Implementing Universal Basic Income (UBI):**
   - Van Parijs and Vanderborght (2017) explore the potential of Universal Basic Income as a solution to technological unemployment, offering a safety net for those displaced by automation.
   - Standing (2017) argues for UBI as a means to reduce inequality and provide economic stability in the face of job displacement, highlighting its role in promoting social cohesion.

3. **Incentivizing Job Creation in Emerging Industries:**
   - Mazzucato (2015) discusses the role of government in driving innovation and job creation in new industries, suggesting targeted investments in sectors likely to expand due to technological advancements.
   - Moretti (2012) highlights the importance of incentivizing private sector investment in high-tech industries, which can lead to job creation and economic revitalization.

4. **Reforming Education Systems:**
   - Goldin and Katz (2008) argue for the reform of education systems to better align with the demands of a technology-driven economy, focusing on STEM education and critical thinking skills.
   - Brynjolfsson and McAfee (2014) stress the need for educational reforms that emphasize creativity, problem-solving, and human-centric skills, which are less susceptible to automation.

5. **Regulatory Frameworks for New Technologies:**
   - West (2018) addresses the need for updated regulatory frameworks that can keep pace with rapid technological

changes, ensuring ethical and responsible use of AI and robotics.

- ○ Lee and Fung (2019) propose the development of international standards and regulations for AI and automation, to address global challenges of workforce displacement and ensure fair competition.

9

Chapter 4: New Horizons: Jobs of the Future

The landscape of the future job market, influenced by rapid technological advancements and changing societal needs, presents a spectrum of new and evolving career paths. This expanded analysis, supported by academic sources, provides a deeper insight into the nature and scope of these emerging jobs.

1. **AI and Machine Learning Specialists**:

The evolution of the job market with the advent of AI and machine learning (ML) heralds a new era of specialized careers. AI and ML specialists are at the forefront of this transformation, playing a pivotal role in shaping various industries. This expanded analysis delves deeper into the significance and future trajectory of these roles, supported by academic references.

1. **Integration Across Industries**:
   - Bughin et al. (2018) highlight the versatility of AI and ML specialists in adapting their skills across sectors like

healthcare, finance, and manufacturing, driving innovation and efficiency.

- Agrawal, Gans, and Goldfarb (2018) discuss how AI specialists are revolutionizing traditional industries by implementing intelligent systems that enhance productivity and decision-making processes.

2. **Role in Advancing Technology**:

- Brynjolfsson and McAfee (2014) emphasize the role of AI and ML specialists in pushing the boundaries of what machines can do, extending their capabilities beyond routine tasks to more complex problem-solving.
- Jordan and Mitchell (2015) explore the ongoing research in machine learning, underscoring the specialists' contribution to developing more advanced and autonomous AI systems.

3. **Ethical and Social Responsibilities**:

- Floridi et al. (2018) delve into the ethical implications of AI development, suggesting that AI specialists have a responsibility to ensure the ethical deployment of AI technologies.
- Crawford and Calo (2016) discuss the societal impacts of AI, highlighting the need for AI and ML specialists to be aware of and address potential biases, privacy concerns, and inequalities that AI systems might perpetuate.

4. **Future Skill Requirements**:

- Kaplan (2016) points out the need for continuous learning and adaptation among AI and ML specialists, as the field is rapidly evolving with new technologies and methodologies.
- Daugherty and Wilson (2018) argue that future AI and ML specialists will require a blend of technical skills and soft skills, such as creativity and emotional intelligence, to effectively integrate AI solutions in human-centric fields.

5. **Job Market Demand**:
    - Manyika et al. (2017) predict a significant increase in demand for AI and ML specialists, driven by the growing adoption of AI technologies across industries.
    - Schwab (2016) envisions a future where AI and ML expertise is not just desirable but essential for businesses looking to remain competitive and innovative.

II. Data Analysts and Scientists:

The increasing reliance on big data across various industries has elevated the importance of data analysts and scientists. This expansion explores their evolving role, the skills required, and the future job market landscape for these professions, supported by relevant academic sources.

1. **Evolving Role in Decision-Making**:
    - Davenport and Patil (2012) emphasize the role of data scientists in extracting meaningful insights from large datasets, transforming how organizations make strategic decisions.
    - Provost and Fawcett (2013) discuss the increasing reliance on data analytics for predictive modeling in sectors like marketing, finance, and healthcare, underscoring the growing importance of these roles.
2. **Skills and Qualifications**:
    - Mayer-Schönberger and Cukier (2013) highlight the need for a combination of statistical, programming, and analytical skills in data science, along with domain-specific knowledge to effectively interpret data.
    - Foreman (2013) stresses the importance of communication skills for data analysts and scientists, enabling them to convey complex findings to non-technical stakeholders.
3. **Impact of AI and Machine Learning**:

- ○ Jordan and Mitchell (2015) point out that the integration of AI and machine learning techniques is becoming essential for data professionals, enhancing their ability to handle large-scale, complex datasets.
- ○ Agrawal, Gans, and Goldfarb (2018) discuss the synergy between data professionals and AI technologies, where AI assists in data processing and analysis, while human judgment is crucial for context and decision-making.

4. **Job Market Trends**:

- ○ Manyika et al. (2017) predict a surge in demand for data analysts and scientists, driven by the exponential growth in data generation and the need for data-driven decision-making in business and governance.
- ○ Bughin et al. (2018) suggest that the data analyst and scientist roles will evolve with technological advancements, requiring continuous learning and adaptation to stay relevant in the job market.

5. **Ethical and Privacy Considerations**:

- ○ Boyd and Crawford (2012) raise concerns about privacy and ethical use of data, emphasizing the responsibility of data professionals in upholding data integrity and ethical standards.
- ○ O'Neil (2016) warns about the potential biases in data analysis and the need for data professionals to be aware of and mitigate these biases to ensure fairness and accuracy in their work.

III. Renewable Energy Technicians:

The shift towards sustainable energy sources has led to a growing demand for renewable energy technicians. This expanded analysis delves into the role of these technicians, the skills required, and the future job prospects in this field, supported by relevant academic and industry sources.

1. **Role and Responsibilities**:
   - Sovacool and Griffiths (2020) describe renewable energy technicians as crucial for the installation, maintenance, and repair of renewable energy systems, including solar, wind, and geothermal technologies.
   - Wiser et al. (2016) highlight the role of these technicians in ensuring the efficiency and reliability of renewable energy systems, which is vital for the energy transition.

2. **Skills and Training**:
   - Lewis (2014) emphasizes the need for a strong foundation in electrical and mechanical systems, as well as specialized training in renewable energy technologies.
   - The International Renewable Energy Agency (IRENA) (2018) reports on the importance of continuous education and training, including safety protocols and emerging technologies in the renewable sector.

3. **Job Market Growth**:
   - The U.S. Bureau of Labor Statistics (2020) predicts a significant growth in employment opportunities for renewable energy technicians, especially in solar and wind energy, driven by government policies and decreasing costs of renewable technologies.
   - Deloitte (2019) suggests that the expansion of renewable energy infrastructure worldwide will create numerous job opportunities in both urban and rural areas.

4. **Technological Advancements and Challenges**:
   - Nygaard and Wieczorek (2018) discuss how advancements in renewable energy technologies, such as improved energy storage systems, will require technicians to acquire new skills and adapt to changing job requirements.
   - Kerr and Walrath (2017) address the challenges posed by the intermittent nature of renewable energy sources, and

the need for technicians to work on integrating renewable systems with existing power grids.

5. **Environmental and Economic Impact**:
   - Jacobson et al. (2017) argue that the growth in renewable energy jobs, including technicians, will have a significant positive impact on reducing carbon emissions and combating climate change.
   - The World Bank (2020) highlights the economic benefits of investing in renewable energy jobs, including local job creation and sustainable economic development.

IV. Healthcare Innovators:

In the ever-evolving landscape of healthcare, the role of healthcare innovators has become increasingly significant. This expanded analysis explores the nature of this role, the skills required, and future prospects, supported by academic and industry sources.

1. **Role and Responsibilities**:
   - According to Shaw, Rudzicz, Jamieson, and Goldfarb (2018), healthcare innovators are pivotal in integrating technological advancements into healthcare practices, focusing on improving patient outcomes and healthcare efficiency.
   - A study by Bhatt (2019) emphasizes their role in developing new healthcare models, digital health solutions, and personalized medicine approaches.

2. **Skills and Training**:
   - Densen (2011) highlights that healthcare innovators require a combination of medical knowledge, technological proficiency, and skills in entrepreneurship.
   - The World Health Organization (WHO) (2017) notes the importance of interdisciplinary training, including aspects of data science, AI, and patient care.

3. **Job Market Growth**:

    ○ The American Medical Association (2020) projects a growing demand for healthcare innovators due to an aging population and the increasing prevalence of chronic diseases.

    ○ Deloitte's Global Healthcare Outlook (2021) forecasts a surge in investment in healthcare innovation, driven by digital health trends and the aftermath of the COVID-19 pandemic.

4. **Technological Advancements and Challenges**:

    ○ A report by Topol (2019) discusses how advancements in genomics, AI, and mobile health technologies present new opportunities and challenges for healthcare innovators.

    ○ Research by Agarwal et al. (2020) addresses the ethical and regulatory challenges in healthcare innovation, especially concerning patient data privacy and AI algorithms.

5. **Impact on Patient Care and Healthcare Systems**:

    ○ Porter (2010) argues that healthcare innovators play a critical role in value-based healthcare, focusing on maximizing patient outcomes relative to costs.

    ○ The McKinsey Global Institute (2018) suggests that innovations in healthcare can significantly reduce costs, improve access, and enhance the quality of care.

V. Urban Farmers and Food Engineers:

The roles of urban farmers and food engineers are becoming increasingly significant in addressing global food security challenges. This analysis delves into the scope, necessary skills, and future prospects of these professions, supported by pertinent literature and industry reports.

1. **Role and Responsibilities**:

- Urban farmers, as described by Orsini, Kahane, Nono-Womdim, and Gianquinto (2013), are responsible for producing food in urban environments, using techniques like vertical farming and hydroponics to maximize space efficiency.
- Food engineers, according to Van Huis and Oonincx (2017), focus on developing sustainable food processing techniques, alternative protein sources, and enhancing food nutrition and safety.

2. **Skills and Training:**
   - Specht et al. (2014) emphasize the need for urban farmers to possess skills in sustainable agriculture practices, urban planning, and community engagement.
   - A study by Costa-Font, Gil, and Traill (2008) highlights that food engineers require knowledge in biotechnology, food science, and engineering principles, along with innovation and problem-solving skills.

3. **Job Market Growth:**
   - The Food and Agriculture Organization of the United Nations (FAO) (2019) predicts an increasing demand for urban farming initiatives to support growing urban populations and promote food security.
   - The Bureau of Labor Statistics (2020) projects growth in the field of food engineering, driven by the need for efficient food production methods and sustainable food systems.

4. **Technological Advancements and Challenges:**
   - Thomaier et al. (2015) discuss the integration of advanced technologies like IoT and AI in urban farming for optimizing crop yields and resource management.
   - A report by Godfray et al. (2010) addresses challenges in food engineering, including adapting to climate change and meeting the nutritional needs of a growing population.

5. **Impact on Society and Environment**:
   - Ackerman (2016) argues that urban farming can significantly contribute to urban biodiversity, waste reduction, and community well-being.
   - A study by Gustavsson, Cederberg, and Sonesson (2011) suggests that innovative food engineering practices can substantially reduce food waste and greenhouse gas emissions, thereby enhancing environmental sustainability.

VI. Cybersecurity Experts:

The role of cybersecurity experts is increasingly crucial in the digital age, where data breaches and cyber threats are prevalent. This expansion focuses on the evolving scope, necessary skills, and future trends of cybersecurity professions, supported by scholarly sources and industry reports.

1. **Role and Responsibilities**:
   - Cybersecurity experts, as Kshetri (2010) notes, are primarily responsible for protecting information systems from cyber threats, including malware, phishing, and network attacks.
   - According to Hadnagy and Fincher (2015), these experts also play a key role in educating employees about security protocols and identifying potential internal vulnerabilities.
2. **Skills and Training**:
   - A study by Leukfeldt, Holt, and Bernaards (2017) highlights that cybersecurity experts must have a strong foundation in computer science, network security, and encryption technologies.
   - Furnell (2014) emphasizes the importance of continuous learning and adaptation, as cybersecurity is a rapidly evolving field with new threats emerging constantly.
3. **Job Market Growth**:

- The Bureau of Labor Statistics (2020) projects a significant increase in demand for cybersecurity professionals, driven by the growing prevalence of cyber threats and data breaches.
- Anderson and Rainie (2017) suggest that the expansion of the Internet of Things (IoT) will further boost the need for cybersecurity expertise.

4. **Challenges and Ethical Considerations**:
- A report by Libicki, Senty, and Pollak (2014) discusses the challenges cybersecurity experts face, including the sophistication of cybercriminals and the complexity of digital infrastructures.
- Taddeo (2016) addresses ethical considerations, stressing the need for cybersecurity experts to balance privacy concerns with security measures.

5. **Impact on Business and Society**:
- Romanosky (2016) explores the economic impacts of cyber incidents, indicating the crucial role of cybersecurity in protecting businesses from financial losses.
- Wall (2007) discusses the broader societal implications, noting that effective cybersecurity is integral to maintaining public trust in digital systems and safeguarding personal data.

Forecasting the AI-driven job market:

The future job market, reshaped by artificial intelligence (AI), is a topic of significant interest and research. This expansion will delve into the transformations expected in various sectors, skills in demand, and the socioeconomic impacts, backed by academic and industry insights.

1. **Sectoral Transformations**:

The impact of Artificial Intelligence (AI) on various sectors is profound and diverse. This expansion focuses on the sectoral transformations due to AI advancements, supported by academic and industry insights.

1. **Healthcare**:
   - In healthcare, AI is revolutionizing patient care and medical diagnostics. Jiang et al. (2017) discuss AI's role in enhancing diagnostic accuracy and personalized medicine. AI technologies are predicted to streamline administrative processes, improve patient outcomes, and assist in complex surgeries.
   - According to Esteva et al. (2019), AI in dermatology and radiology shows promise in early disease detection, significantly impacting patient care.

2. **Finance**:
   - AI's impact on the finance sector includes algorithmic trading, risk management, and personalized banking services. Bao et al. (2021) explore how AI algorithms have transformed trading strategies, leading to more efficient financial markets.
   - Arner et al. (2017) highlight AI's role in fraud detection and customer service enhancement in banking, suggesting a shift towards more secure and customer-centric financial services.

3. **Manufacturing**:
   - In manufacturing, AI optimizes production lines, reduces downtime, and improves quality control. Lee et al. (2018) describe the integration of AI in smart manufacturing, which enhances predictive maintenance and supply chain management.
   - Rüßmann et al. (2015) argue that AI-driven automation in manufacturing not only increases efficiency but also

creates new roles, such as AI system managers and data analysts.

4. **Retail**:
   - The retail sector is experiencing a transformation with AI in inventory management, customer experience, and personalized marketing. Huang and Rust (2018) discuss AI's role in creating personalized shopping experiences, improving customer engagement and sales.
   - Grewal et al. (2020) emphasize AI's impact on supply chain optimization and predictive analytics, leading to more efficient stock management and customer satisfaction.

5. **Education**:
   - AI in education is reshaping learning experiences through personalized learning and automated administrative tasks. Luckin et al. (2016) explore AI's potential in creating adaptive learning environments, catering to individual student needs.
   - Zhou et al. (2020) highlight AI's role in automating administrative tasks, allowing educators to focus more on teaching and less on paperwork.

6. **Transportation and Logistics**:
   - AI significantly impacts transportation and logistics through autonomous vehicles and route optimization. Litman (2020) discusses the potential of self-driving cars in reducing accidents and improving traffic management.
   - Savelsbergh and Van Woensel (2016) explore AI's application in logistics for efficient route planning and delivery, leading to cost savings and environmental benefits.

II. Skills in Demand:

The AI-driven job market of the future will require a unique set of skills, emphasizing technical proficiency, creativity, and adaptability.

This section provides a deeper analysis of these skills, supported by scholarly sources.

1. **Technical and Analytical Skills**:
   - Proficiency in data analysis, machine learning, and programming will be essential. Bessen (2019) asserts the growing importance of digital literacy in the AI era, stressing the need for skills in data science and coding. The ability to understand and manipulate large data sets will be crucial across various sectors.
   - Bughin, Hazan, and Ramaswamy (2018) emphasize the need for skills in machine learning and artificial intelligence, as these are the core competencies driving the AI revolution.

2. **Problem-Solving and Critical Thinking**:
   - AI-enhanced workplaces will demand advanced problem-solving abilities. Davenport and Kirby (2016) highlight the need for critical thinking in interpreting AI outputs and making strategic decisions.
   - Skills in logical reasoning and complex problem-solving will become more valuable as AI systems handle routine tasks, leaving more intricate problems to humans.

3. **Creativity and Innovation**:
   - As AI automates routine tasks, creative skills will be highly sought after. Florida (2019) notes that creativity will be indispensable in the AI-driven economy, as it cannot be easily automated.
   - Pink (2019) echoes this sentiment, suggesting that jobs requiring creativity, empathy, and innovation will see growth, as these are areas where AI cannot easily replicate human capabilities.

4. **Adaptability and Lifelong Learning**:

- The rapid evolution of technology will necessitate continual learning and adaptability. Schwartz et al. (2020) argue that the ability to adapt and continuously update one's skill set will be essential in the AI-driven job market.
- This includes not just learning new technologies but also adapting to evolving workplace cultures and practices influenced by AI integration.

5. **Interpersonal and Emotional Intelligence**:

- Emotional intelligence will be critical in professions where human interaction is key. Goleman (2017) highlights the importance of empathy, social skills, and emotional intelligence in the future workplace.
- As AI handles more analytical tasks, human-centric skills in communication, collaboration, and leadership will become increasingly important.

III. Socioeconomic Impacts:

The AI-driven job market will not only transform the nature of work but also have significant socioeconomic impacts. This section delves deeper into these impacts, drawing on relevant literature.

1. **Income Inequality**:

- The rise of AI could exacerbate income inequality. Brynjolfsson and McAfee (2014) argue that while AI boosts productivity and generates wealth, its benefits are not evenly distributed, often favoring those with high-skill levels and access to technology.
- Autor (2015) discusses how technological advancements may widen the gap between high-wage, high-skill jobs and low-wage, low-skill jobs, intensifying economic disparities.

2. **Job Market Polarization**:

- AI and automation can lead to job market polarization, where mid-level jobs are eroded while low-skill and high-skill jobs expand. Goos, Manning, and Salomons (2014) describe this phenomenon, noting the diminishing number of routine, middle-skill jobs.
- This polarization can lead to social and economic challenges, as workers struggle to adapt to changing job requirements.

3. **Workforce Displacement and Re-skilling Challenges**:
   - The displacement of workers due to AI integration poses a significant challenge. Frey and Osborne (2017) predict that up to 47% of U.S. jobs are at risk of automation, necessitating large-scale re-skilling programs.
   - The re-skilling challenge is significant, as noted by the World Economic Forum (2018), which emphasizes the need for lifelong learning and upskilling to keep pace with technological changes.

4. **Changes in Labor Demand**:
   - AI will shift labor demand towards more tech-oriented and human-centered roles. Acemoglu and Restrepo (2019) argue that while AI displaces some jobs, it also creates new ones, often in areas that are complementary to AI technologies.
   - The demand for soft skills, like emotional intelligence, is likely to increase, as these are areas where AI cannot easily replicate human capabilities (Brynjolfsson et al., 2019).

5. **Public Policy and Social Welfare**:
   - AI's impact on the job market will necessitate changes in public policy, especially in areas of education, labor, and social welfare. Susskind and Susskind (2015) suggest that governments need to rethink social welfare systems to support workers displaced by AI.

- Policies focusing on universal basic income, job retraining programs, and enhanced education systems are discussed as potential responses to the challenges posed by AI (Yang, 2018).

IV. Educational Implications:

The advent of AI in the job market necessitates a reevaluation of educational systems to prepare future generations for new challenges and opportunities. This expanded analysis delves into the educational implications of an AI-driven job market.

1. **Shift in Educational Focus:**
   - A shift towards STEM (Science, Technology, Engineering, Mathematics) and digital literacy is crucial. The World Economic Forum (2020) emphasizes the growing need for proficiency in digital skills alongside traditional academic knowledge.
   - Schwab (2016) argues for integrating AI and coding into curricula, preparing students for a technology-centric job market.

2. **Emphasis on Soft Skills:**
   - Alongside technical skills, the importance of soft skills such as critical thinking, creativity, and emotional intelligence is accentuated (Brynjolfsson et al., 2018). These skills enable adaptation in a rapidly evolving job landscape where human-centric roles gain prominence.
   - The development of social skills, problem-solving abilities, and adaptability is essential, as noted by Deming (2017), to complement AI and automation.

3. **Lifelong Learning and Continuous Education:**
   - The concept of lifelong learning becomes indispensable in the AI era. Bughin et al. (2018) stress the importance of

continuous learning and upskilling to stay relevant in the job market.

- Educational systems need to provide platforms for ongoing skill development, moving beyond traditional schooling to continuous, flexible education models.

4. **Customized and Adaptive Learning**:

- AI in education can enable personalized learning experiences. Luckin et al. (2016) discuss how AI can adapt to individual learning styles, pace, and preferences, enhancing educational outcomes.
- This personalized approach can better prepare students for specific career paths influenced by AI and technology advancements.

5. **Ethics and AI Education**:

- Ethical education surrounding AI becomes increasingly important. Educating students about the ethical implications of AI, as discussed by Floridi et al. (2018), is vital to develop responsible AI usage and understanding.
- Courses on AI ethics, data privacy, and the societal impact of technology should be integral parts of education curricula.

6. **Partnerships between Education and Industry**:

- Collaboration between educational institutions and industry is necessary to align educational programs with market needs. This partnership can ensure curriculum relevance and provide practical exposure to AI applications in various fields (Agrawal et al., 2018).

V. Ethical and Policy Considerations:

The integration of AI into the job market brings to the fore various ethical and policy considerations that need addressing. This expanded analysis examines these aspects in greater depth.

1. **Data Privacy and Security**:
   - As AI systems rely heavily on data, concerns about data privacy and security are paramount. The General Data Protection Regulation (GDPR) sets a precedent for how personal data should be handled (European Commission, 2018).
   - Polonski (2017) discusses the need for robust data protection policies to safeguard against unauthorized data access and breaches.

2. **Bias and Fairness in AI**:
   - AI algorithms can perpetuate existing biases if not carefully designed. Angwin et al. (2016) highlighted instances where AI systems showed racial and gender biases.
   - Policies must enforce fairness in AI, ensuring algorithms are transparent and accountable, as argued by Zou and Schiebinger (2018).

3. **Impact on Employment and Inequality**:
   - AI could exacerbate economic inequality by displacing jobs in certain sectors. Acemoglu and Restrepo (2020) explore the differential impact of automation on various demographic groups.
   - Policies should address retraining and support for workers displaced by AI, as suggested by Susskind (2020), to mitigate income inequality.

4. **Ethical AI Development and Use**:
   - The ethical development and deployment of AI is crucial. Floridi et al. (2018) discuss the need for ethical guidelines in AI to ensure its beneficial and non-malicious use.
   - Government and industry standards must be established to guide ethical AI practices, including transparency in AI decision-making processes.

5. **Regulatory Frameworks and Standardization**:

- A comprehensive regulatory framework for AI is necessary. As Bostrom and Yudkowsky (2014) point out, regulations should balance innovation with safety and ethical considerations.
- Standardization of AI technologies, as recommended by IEEE (2019), would help in creating global norms and best practices.

6. **Global Cooperation and Policy Coordination**:
   - International collaboration is essential to address the global implications of AI. The OECD's Principles on AI (2019) represent a multinational effort to promote AI that is innovative and trustworthy.
   - Coordinated policy efforts are needed to address cross-border challenges such as intellectual property rights and international labor market impacts.

Emerging roles and industries in the AI epoch:

The AI epoch is not just transforming existing jobs but also creating entirely new roles and industries. This expanded analysis focuses on these emerging trends, supported by relevant research and data.

1. **AI Ethics Officer:**

In the rapidly evolving landscape of artificial intelligence (AI), the position of an AI Ethics Officer is becoming increasingly crucial. This role focuses on navigating the complex ethical, legal, and social implications of AI technologies.

1. **Defining the Role of AI Ethics Officer**:
   - The AI Ethics Officer is responsible for developing and implementing ethical guidelines and standards for AI usage within organizations. As outlined by Jobin, Ienca, and

Vayena (2019), these professionals play a pivotal role in ensuring that AI systems are designed and deployed in a manner that respects human rights, privacy, and fairness.

- Their duties include conducting ethical audits of AI systems, advising on ethical AI design, and ensuring compliance with regulatory and legal standards related to AI (Hagendorff, 2020).

2. **Importance in the Business and Technological Context**:

- As AI becomes more integrated into business processes, the AI Ethics Officer becomes vital in bridging the gap between technology and ethical considerations. According to Martin (2019), these officers ensure that AI systems align with both organizational values and societal norms.
- Their role also extends to risk management, where they assess potential ethical risks associated with AI technologies and propose mitigation strategies (Martin, 2019).

3. **Challenges and Responsibilities**:

- One of the primary challenges faced by AI Ethics Officers is keeping pace with the rapid development of AI technologies and their diverse applications. Hagendorff (2020) emphasizes the dynamic nature of this field, where ethical guidelines must continuously evolve to address new challenges.
- AI Ethics Officers are also responsible for fostering an organizational culture that values ethical considerations in AI, which includes training and awareness programs for employees (Mittelstadt, 2019).

4. **Future Prospects and Development**:

- The demand for AI Ethics Officers is expected to grow as more organizations recognize the importance of ethical considerations in AI. Ryan and Stahl (2020) predict an increase in demand for professionals with expertise in AI ethics across various sectors.

- The role is also likely to evolve, with AI Ethics Officers playing a more prominent role in strategic decision-making within organizations (Ryan & Stahl, 2020).

II. AI-Assisted Healthcare Professional:

The emergence of AI in healthcare has led to the evolution of a new role: the AI-Assisted Healthcare Professional. This role represents a synergy between human medical expertise and AI technologies, aiming to enhance patient care and medical outcomes.

1. **Role and Responsibilities**:
   - AI-Assisted Healthcare Professionals utilize AI tools for diagnostics, treatment planning, and patient monitoring. According to Jiang et al. (2017), these professionals integrate AI insights into clinical decision-making, improving the accuracy and efficiency of diagnoses and treatments.
   - They are also responsible for managing and interpreting data from AI systems, ensuring that the information is used effectively in patient care (Topol, 2019).

2. **Impact on Patient Care**:
   - The incorporation of AI in healthcare has significantly improved patient outcomes. As Davenport and Kalakota (2019) note, AI-assisted tools can detect diseases earlier and with greater precision, leading to more effective treatments.
   - These professionals also personalize patient care by leveraging AI's predictive analytics, thus enhancing the quality of treatment (Davenport & Kalakota, 2019).

3. **Educational and Skill Requirements**:
   - This role requires a blend of medical knowledge and proficiency in AI technologies. Blease et al. (2019) emphasize the need for continuous education in both medical and AI fields to stay abreast of evolving technologies.

- Skills in data analysis, ethical application of AI in healthcare, and patient communication are also essential (Luxton, 2020).

4. **Challenges and Ethical Considerations**:
- AI-Assisted Healthcare Professionals face challenges in ensuring the ethical use of AI, particularly regarding patient privacy and data security (Luxton, 2020).
- They must also navigate the limitations of AI tools, ensuring that AI supplements but does not replace human judgment in clinical settings (Topol, 2019).

5. **Future Outlook**:
- The demand for AI-Assisted Healthcare Professionals is expected to grow as AI technologies become more integrated into healthcare systems. Blease et al. (2019) predict an increasing need for professionals who can effectively combine medical expertise with AI skills.
- The role is likely to evolve with advancements in AI, potentially leading to new specialties within healthcare focused on AI applications (Jiang et al., 2017).

II. Renewable Energy Technicians and Engineers:

The integration of AI into the renewable energy sector is fostering the emergence of specialized roles for Renewable Energy Technicians and Engineers. These professionals are pivotal in optimizing and maintaining renewable energy systems, utilizing AI for greater efficiency and sustainability.

1. **Role and Responsibilities**:
- Renewable Energy Technicians and Engineers are tasked with designing, implementing, and maintaining renewable energy systems (solar, wind, etc.), often using AI algorithms for system optimization (Dincer & Rosen, 2020).

- They use AI to analyze energy patterns, predict maintenance needs, and optimize energy production, ensuring maximum efficiency and reliability (Kalogirou, 2018).

2. **Impact on Renewable Energy Systems**:
   - The application of AI in renewable energy enhances the capacity to predict energy outputs, leading to more stable and efficient energy grids (Zhang et al., 2018).
   - As noted by Li et al. (2019), AI-driven analytics help in the early detection of system faults, reducing downtime and maintenance costs.

3. **Educational and Skill Requirements**:
   - This role demands a strong foundation in renewable energy technologies, coupled with skills in AI and data analysis (Dincer & Rosen, 2020).
   - Continuous learning is crucial, as the field is rapidly evolving with new AI applications and renewable technologies (Kalogirou, 2018).

4. **Challenges and Future Outlook**:
   - One of the primary challenges is keeping pace with the fast-evolving AI technologies and their applications in renewable energy (Li et al., 2019).
   - The future outlook is promising, with growing demand for these professionals as the world shifts towards sustainable energy sources. The role is expected to evolve with advancements in both renewable technologies and AI (Zhang et al., 2018).

5. **Economic and Environmental Implications**:
   - The advancement in AI-assisted renewable energy technologies is likely to drive down the costs of renewable energy, making it more accessible and competitive with traditional energy sources (Kalogirou, 2018).

- This shift has significant environmental benefits, contributing to reduced carbon emissions and a more sustainable future (Dincer & Rosen, 2020).

III. Urban Mobility Managers:

In the AI-driven epoch, the role of Urban Mobility Managers is emerging as a critical one in shaping the future of urban transportation. This role involves leveraging AI and technology to develop, implement, and manage efficient, sustainable, and intelligent urban mobility solutions.

1. **Role and Responsibilities**:
   - Urban Mobility Managers are responsible for integrating AI technologies into urban transportation systems, including autonomous vehicles, intelligent traffic management, and smart public transit systems (Schrank et al., 2019).
   - They focus on optimizing traffic flow, reducing congestion, and enhancing the safety and accessibility of urban transportation (Litman, 2020).

2. **Impact on Urban Transportation**:
   - The application of AI in urban mobility leads to more efficient and adaptive traffic management systems, reducing travel time and environmental impact (Zhao et al., 2019).
   - AI-driven analytics enable better planning and forecasting of transportation needs, improving the overall user experience (Litman, 2020).

3. **Educational and Skill Requirements**:
   - Urban Mobility Managers typically require a background in urban planning or transportation engineering, combined with expertise in AI, data analysis, and smart technology applications (Schrank et al., 2019).

- Skills in project management, policy development, and stakeholder engagement are also essential (Zhao et al., 2019).

4. **Challenges and Future Outlook**:
   - Challenges include integrating various modes of transportation within smart city frameworks and addressing privacy and security concerns related to data collection and usage (Litman, 2020).
   - The future outlook suggests a growing demand for this role, with increasing urbanization and the need for sustainable, efficient urban mobility solutions (Zhao et al., 2019).

5. **Economic and Environmental Implications**:
   - Effective urban mobility management can lead to significant economic benefits by reducing transportation costs and improving productivity through reduced travel times (Schrank et al., 2019).
   - Environmentally, it contributes to reduced greenhouse gas emissions and pollution, aligning with sustainable urban development goals (Litman, 2020).

IV. **AI-Driven Supply Chain Managers**:

As the landscape of supply chain management transforms with advancing technology, the emergence of AI-Driven Supply Chain Managers marks a significant shift in this sector. This role revolves around integrating artificial intelligence to enhance efficiency, predictability, and responsiveness in supply chain operations.

1. **Role and Responsibilities**:
   - AI-Driven Supply Chain Managers are tasked with implementing AI systems to streamline supply chain processes, including inventory management, logistics, demand forecasting, and supplier relationship management (Ivanov, Dolgui, & Sokolov, 2019).

- They focus on utilizing data-driven insights to optimize supply chain operations, reduce costs, and improve customer satisfaction (Fosso Wamba & Queiroz, 2020).

2. **Impact on Supply Chain Operations**:
- The integration of AI allows for real-time analytics and decision-making, enhancing the agility and resilience of supply chains (Ivanov et al., 2019).
- AI enables predictive analytics for demand forecasting, reducing overstock and stockouts, and facilitating just-in-time inventory management (Fosso Wamba & Queiroz, 2020).

3. **Educational and Skill Requirements**:
- This role typically requires a background in supply chain management or logistics, combined with knowledge in AI, machine learning, and data analytics (Ivanov et al., 2019).
- Skills in strategic planning, technological aptitude, and change management are essential (Fosso Wamba & Queiroz, 2020).

4. **Challenges and Future Outlook**:
- Challenges include ensuring data security and privacy, managing the ethical implications of AI, and adapting to rapidly evolving technology (Ivanov et al., 2019).
- The future outlook indicates an increasing reliance on AI in supply chain management, necessitating a continuous upskilling of managers in this field (Fosso Wamba & Queiroz, 2020).

5. **Economic and Efficiency Implications**:
- AI-driven supply chain management can significantly reduce operational costs and increase efficiency through automation and optimized resource allocation (Ivanov et al., 2019).

- It contributes to enhanced competitiveness and sustainability in the global market (Fosso Wamba & Queiroz, 2020).

V. **Digital Transformation Specialists**:

The role of Digital Transformation Specialists has become increasingly pivotal in the era of rapid technological advancement. These professionals are integral to guiding organizations through the complexities of integrating digital technologies, including AI, into their business models and operational processes.

1. **Role and Responsibilities**:
   - Digital Transformation Specialists lead the strategic planning and execution of digital initiatives within organizations to enhance operational efficiency, customer experience, and innovation (Verhoef, Broekhuizen, Bart, Bhattacharya, Dong, Fabian, & Haenlein, 2021).
   - They collaborate with various departments to align digital strategies with business objectives, ensuring a cohesive transformation across the organization (Hess, Matt, Benlian, & Wiesböck, 2016).
2. **Impact on Business Operations**:
   - By implementing digital solutions, these specialists can significantly improve business processes, customer engagement, and data-driven decision-making (Verhoef et al., 2021).
   - Digital transformation enhances an organization's agility, enabling it to adapt quickly to market changes and emerging trends (Hess et al., 2016).
3. **Educational and Skill Requirements**:
   - This role typically requires a background in business management, information technology, or a related field,

combined with a deep understanding of digital technologies and their business applications (Hess et al., 2016).

- Skills in project management, strategic planning, and leadership are crucial, along with the ability to navigate the cultural changes associated with digital transformation (Verhoef et al., 2021).

4. **Challenges and Future Outlook**:

- Challenges include managing the resistance to change within organizations, ensuring data security, and staying abreast of rapidly evolving technologies (Hess et al., 2016).
- The future outlook suggests a continuous demand for these specialists as digital transformation becomes a necessity for competitive advantage in various industries (Verhoef et al., 2021).

5. **Economic and Competitive Implications**:

- Effective digital transformation can lead to significant cost savings, revenue growth, and improved customer satisfaction (Hess et al., 2016).
- It provides a competitive edge by enabling organizations to innovate, personalize customer experiences, and optimize operations (Verhoef et al., 2021).

VI. **Sustainable Agriculture Technologists**:

Sustainable Agriculture Technologists are becoming crucial in modern agriculture, particularly in the context of integrating AI and other advanced technologies to promote sustainable farming practices. Their role is central to addressing the increasing challenges of food security, environmental sustainability, and climate change.

1. **Role and Responsibilities**:

- These professionals focus on developing and implementing technological solutions that enhance agricultural efficiency while minimizing environmental impact (Pretty, Toulmin, & Williams, 2011).

- They are involved in deploying AI-driven tools like precision farming, which involves using data analytics for optimizing planting, watering, and harvesting (Liakos, Busato, Moshou, Pearson, & Bochtis, 2018).

2. **Impact on Agricultural Practices**:

- AI technologies enable more precise resource management, reducing waste and improving yield (Pretty et al., 2011).
- Sustainable Agriculture Technologists contribute to the development of environmentally friendly farming methods, such as integrating renewable energy sources into agricultural practices (Liakos et al., 2018).

3. **Educational and Skill Requirements**:

- A background in agricultural sciences, along with knowledge in environmental science and technology, is essential (Liakos et al., 2018).
- Skills in data analytics, AI applications, and a strong understanding of sustainable agricultural practices are required (Pretty et al., 2011).

4. **Challenges and Future Outlook**:

- Key challenges include adapting to rapidly changing technology, addressing the digital divide in rural areas, and managing the transition towards sustainable practices (Pretty et al., 2011).
- The future outlook suggests a growing demand for these roles as global food systems face the dual challenge of increasing productivity and reducing environmental impact (Liakos et al., 2018).

5. **Economic and Environmental Implications**:

- ○ Sustainable agriculture technologies can lead to cost savings for farmers, higher yields, and more efficient use of resources (Pretty et al., 2011).
- ○ Environmentally, these technologies offer potential solutions for reducing greenhouse gas emissions and conserving biodiversity (Liakos et al., 2018).

Preparing today's workforce for tomorrow's jobs:

Preparing the current workforce for future job landscapes is a crucial aspect of economic and social planning, given the rapid technological advancements and changing job requirements. This preparation involves multifaceted strategies and collaborative efforts between educators, policymakers, and industry leaders.

1. **Skills Development and Lifelong Learning**:

Skills development and lifelong learning are integral in preparing today's workforce for future job landscapes, characterized by rapid technological advancements and evolving job requirements.

1. **Importance of Skills Development**:
   - ○ In the context of the Fourth Industrial Revolution, skills development, particularly in areas such as digital literacy, critical thinking, and problem-solving, is essential (Schwab, 2016).
   - ○ The World Economic Forum (2018) emphasizes the growing need for skills like complex problem solving, emotional intelligence, and cognitive flexibility in the future workforce.
2. **Lifelong Learning as a Continuous Process**:

- Lifelong learning is increasingly recognized as a necessity rather than a luxury, ensuring workers remain competitive and adaptable (Bughin et al., 2018).
- The OECD (2019) highlights that continuous learning should be embedded into the culture of workplaces to maintain skill relevance.

3. **Adapting to Technological Changes**:
- Workers need to adapt to the rapidly evolving technological landscape, requiring ongoing education and training (Frey & Osborne, 2017).
- According to the McKinsey Global Institute (2018), around half of today's work activities have the potential to be automated, necessitating new skill sets for employees.

4. **Role of Educational Institutions and Online Platforms**:
- Educational institutions must revise curricula to include more future-oriented skills and competencies (Fadel, 2016).
- Online learning platforms, such as Coursera and Udemy, are playing a pivotal role in making lifelong learning accessible and affordable (Brynjolfsson & McAfee, 2014).

5. **Government and Corporate Involvement**:
- Governments should incentivize lifelong learning through policies, subsidies, and tax breaks (Schwab, 2016).
- Corporations have a responsibility to provide continuous learning opportunities for their employees to stay relevant (Bughin et al., 2018).

6. **Building a Learning Culture**:
- Creating a culture that values and encourages continuous learning is crucial (Schwab, 2016).
- Employers should foster an environment where seeking knowledge and skills is recognized and rewarded (Bughin et al., 2018).

7. **Equity in Access to Learning Opportunities**:

- Efforts must be made to ensure equitable access to learning opportunities for all segments of the population, particularly for underrepresented and disadvantaged groups (OECD, 2019).

II. Educational System Transformation:

The transformation of the educational system is pivotal in equipping today's workforce for the future job market, marked by rapid technological evolution and changing skill demands.

1. **Realigning Education with Future Job Markets**:
   - Educational systems must align more closely with future labor market requirements, emphasizing digital literacy, critical thinking, and soft skills (Schwab, 2016).
   - A study by the World Economic Forum (2018) suggests integrating future-oriented skills and knowledge areas into school curricula.
2. **Focus on STEM Education**:
   - STEM (Science, Technology, Engineering, Mathematics) education is increasingly crucial due to the growing demand for technical skills (Fadel, 2016).
   - The OECD (2019) emphasizes the need for a strong foundation in STEM to prepare students for technological and scientific advancements.
3. **Integration of Technology in Learning**:
   - Integrating technology into education, such as AI and VR, can provide more immersive and personalized learning experiences (Brynjolfsson & McAfee, 2014).
   - The McKinsey Global Institute (2018) highlights the potential of technology-enhanced learning in developing critical skills for the digital age.
4. **Promoting Soft Skills and Lifelong Learning**:

- Soft skills like emotional intelligence, adaptability, and collaboration are becoming as important as technical skills (Bughin et al., 2018).
- Lifelong learning should be encouraged from an early age, fostering a mindset of continuous skill development (Schwab, 2016).

5. **Industry-Academia Collaboration:**

- Partnerships between educational institutions and industries can ensure that curricula stay relevant to evolving job market needs (Frey & Osborne, 2017).
- These collaborations can provide practical experience through internships, co-op programs, and real-world project work (World Economic Forum, 2018).

6. **Educational Equity and Accessibility:**

- Addressing disparities in educational access and quality is vital to ensuring that all students are prepared for future challenges (OECD, 2019).
- Online and remote learning platforms can democratize access to quality education, especially in underserved communities (Brynjolfsson & McAfee, 2014).

7. **Policy Initiatives and Government Support:**

- Government policies need to support educational reforms and innovations that cater to future job requirements (Fadel, 2016).
- Financial support, policy frameworks, and incentives for educational institutions are crucial for these transformations (Schwab, 2016).

III. **Policy Initiatives and Government Role:**

The role of government and policy initiatives is crucial in shaping the future workforce to meet the demands of emerging job markets influenced by technological advancements.

1. **Policy Development for Future Skills:**
   - Governments need to develop policies that focus on equipping the workforce with skills relevant to future job markets (Schwab, 2016).
   - This includes promoting digital literacy, problem-solving, and adaptability skills (World Economic Forum, 2018).

2. **Investment in Educational Infrastructure:**
   - Investment in educational infrastructure is essential to ensure access to quality education and training (OECD, 2019).
   - This includes building and upgrading facilities, providing modern equipment, and ensuring internet connectivity for digital learning platforms.

3. **Subsidizing and Supporting Lifelong Learning:**
   - Governments can subsidize training programs and encourage lifelong learning initiatives (Brynjolfsson & McAfee, 2014).
   - Tax incentives or financial support for individuals pursuing further education and training can also be effective (Bughin et al., 2018).

4. **Public-Private Partnerships:**
   - Collaborations between government, industry, and educational institutions can lead to more effective training programs (Frey & Osborne, 2017).
   - These partnerships can help in aligning curricula with industry needs and providing practical experience through internships and apprenticeships.

5. **Inclusive Education Policies:**
   - Policies must ensure inclusive and equitable quality education to all, reducing disparities based on location, background, or economic status (Fadel, 2016).

- Special attention is needed for marginalized communities to provide them with equal opportunities in the evolving job market (Schwab, 2016).

6. **Labor Market Reforms**:

- Labor market policies need to be reformed to accommodate the changing nature of work, such as the gig economy and remote working (World Economic Forum, 2018).
- Regulations surrounding job security, benefits, and working conditions need to adapt to these new work models.

7. **Anticipating Future Trends**:

- Governments should invest in research and analysis to anticipate future job market trends (OECD, 2019).
- This can help in formulating policies that are proactive rather than reactive to market changes.

8. **Fostering Innovation and Entrepreneurship**:

- Supporting innovation and entrepreneurship through grants, mentorship programs, and incubators can create new job opportunities (Brynjolfsson & McAfee, 2014).
- Policies should encourage risk-taking and innovation, fostering a culture of creativity and growth.

IV. Corporate Responsibility and Reskilling Initiatives:

Corporate responsibility in workforce preparation, particularly through reskilling initiatives, is increasingly recognized as a vital component in addressing the challenges posed by the evolving job market.

1. **Corporate-Led Reskilling Programs**:

- Companies are initiating reskilling programs to address the skill gaps in their workforce, a move that is not only beneficial for employees but also for the long-term viability of the business (Bughin et al., 2018).

- These programs focus on training employees in new technologies, soft skills, and adaptive thinking (World Economic Forum, 2020).

2. **Investment in Employee Development**:
   - Progressive organizations are investing more in their employees' development, acknowledging that continuous learning is essential in a rapidly changing job environment (Schwab, 2016).
   - This includes offering workshops, online courses, and on-the-job training opportunities.

3. **Partnerships with Educational Institutions**:
   - Businesses are increasingly partnering with universities and vocational schools to develop tailored curricula that meet specific industry needs (Frey & Osborne, 2017).
   - These partnerships can provide practical experience and industry-relevant skills to students and existing employees.

4. **Promoting a Culture of Lifelong Learning**:
   - Companies are fostering a culture that values and encourages lifelong learning, recognizing that the ability to learn and adapt is a key skill in the future workforce (OECD, 2019).
   - This can be achieved through mentorship programs, learning incentives, and creating a supportive learning environment within the organization.

5. **Supporting Career Transition and Mobility**:
   - Corporations are providing support for employees transitioning to new roles or careers, particularly those impacted by automation and digital transformation (Brynjolfsson & McAfee, 2014).
   - This includes career counseling, transition programs, and financial support for further education.

6. **Inclusive Training Opportunities**:

- There is an increasing emphasis on providing inclusive training opportunities to ensure diversity in the workplace (Fadel, 2016).
- This involves creating training programs that are accessible to employees of varying backgrounds, abilities, and levels of experience.

7. **Aligning Corporate Strategy with Workforce Development:**

- Forward-thinking companies are aligning their corporate strategy with workforce development (Bughin et al., 2018).
- This ensures that employee skillsets evolve in tandem with the company's long-term goals and market demands.

8. **Responsibility Towards Displaced Workers:**

- Companies are recognizing their responsibility towards workers displaced by technological advancements and are implementing programs to assist them (World Economic Forum, 2020).
- This includes offering retraining programs and assistance in finding new employment opportunities.

V. Anticipating Future Job Trends:

Anticipating future job trends is a crucial aspect of preparing today's workforce for the evolving job market. This proactive approach involves understanding the direction of technological advancements, economic shifts, and the changing nature of work.

1. **Technology-Driven Job Creation:**

- With the rapid advancement of technologies like artificial intelligence, machine learning, and robotics, new job roles are being created, especially in tech-driven sectors (Manyika et al., 2017).

- Identifying these emerging roles and understanding the skills required for them is vital for workforce preparedness (Brynjolfsson & McAfee, 2014).

2. **Sector-Specific Trend Analysis:**

- Different sectors will experience varying impacts from technological advancements. For instance, healthcare, education, and tech sectors are expected to see significant job growth (Frey & Osborne, 2017).
- Understanding these sector-specific trends is essential for guiding workforce development programs (World Economic Forum, 2020).

3. **Impact of Automation on Employment:**

- Automation is likely to replace certain job roles, particularly those involving routine tasks. However, it also creates new opportunities in fields like robotics maintenance and AI supervision (Acemoglu & Restrepo, 2020).
- Preparing the workforce for these changes involves both reskilling for new roles and adapting to working alongside automated systems.

4. **Shift to a Gig Economy:**

- The rise of the gig economy, marked by freelance, contract, and part-time work, is reshaping the employment landscape (De Stefano, 2015).
- Training programs need to include skills for navigating and succeeding in a gig economy, such as entrepreneurship, self-management, and digital literacy.

5. **Globalization and Remote Work Trends:**

- Globalization and the increasing feasibility of remote work are creating a more interconnected and geographically diverse workforce (Baldwin, 2016).
- Adapting to these trends requires cross-cultural communication skills and familiarity with digital collaboration tools.

6. **Sustainability and Green Jobs**:
   - The growing emphasis on sustainability is leading to the creation of 'green jobs' in sectors like renewable energy, sustainable agriculture, and environmental management (Renner, Sweeney, & Kubit, 2008).
   - Training programs must align with this trend, focusing on environmental awareness and skills relevant to the green economy.

7. **Lifelong Learning and Adaptability**:
   - As job trends continue to evolve rapidly, the ability to engage in lifelong learning and adapt to new roles becomes increasingly important (Schwab, 2016).
   - Emphasizing adaptability and continuous learning as key competencies is essential in workforce development.

VI. **Inclusive and Equitable Access**:

Inclusive and equitable access to education and training is a fundamental aspect of preparing today's workforce for future job trends. This approach ensures that all segments of society, regardless of their socio-economic background, gender, or ethnicity, have equal opportunities to develop skills relevant to the emerging job market.

1. **Bridging the Digital Divide**:
   - The digital divide, a significant barrier to equitable access, refers to the gap between individuals with and without access to digital technology and the internet (Van Dijk, 2020).
   - Efforts to bridge this divide are crucial, such as providing widespread internet access and digital literacy training, especially in underprivileged and rural areas.

2. **Gender Equality in Workforce Training**:

- Historically, certain industries, particularly those in STEM fields, have seen underrepresentation of women (Corbett & Hill, 2015).
- Promoting gender equality in these areas through targeted training programs and policies can help balance workforce representation.

3. **Support for Underrepresented Groups:**
- Ethnic minorities, people with disabilities, and other underrepresented groups often face additional barriers to workforce participation (Pager & Shepherd, 2008).
- Initiatives aimed at providing tailored training programs and workplace accommodations are vital for inclusive workforce development.

4. **Affordable and Accessible Education:**
- The high cost of education and training can be a significant barrier to workforce preparation (Deming, 2017).
- Policies that make education more affordable, such as scholarships, grants, and subsidized training programs, are key to ensuring equitable access.

5. **Lifelong Learning Opportunities:**
- Lifelong learning should be accessible to all, regardless of age or current employment status, to adapt to changing job trends (Field, 2000).
- Public and private sectors can collaborate to offer flexible, inclusive, and affordable lifelong learning opportunities.

6. **Cultural Competence and Inclusivity in Training:**
- Training programs must be culturally competent and inclusive, considering the diverse needs and backgrounds of learners (Sue, 2001).
- This approach includes curriculum development that is respectful of and responsive to the cultural contexts of different groups.

7. **Policy Support for Inclusive Workforce Development:**

- Governments play a crucial role in developing policies that promote inclusive workforce development (Bivens et al., 2016).
- Such policies may include anti-discrimination laws, incentives for diversity in hiring, and support for training programs that focus on inclusivity.

10

Chapter 5: The Gig Economy and AI

The intersection of the gig economy and artificial intelligence (AI) is reshaping the nature of work and employment. This shift is characterized by a transition from traditional, long-term employment to more flexible, short-term contracts and freelance work, facilitated by AI technologies.

1. **AI-driven Platforms in the Gig Economy:**

AI-driven platforms are revolutionizing the gig economy by leveraging advanced technologies to enhance the efficiency and effectiveness of how gig work is allocated and performed. These platforms encompass a wide range of sectors, from transportation to freelance services, significantly impacting the labor market.

1. **Optimization of Task Matching:**
    - AI algorithms are adept at matching gig workers with tasks, taking into account factors like skills, location, and availability (Prassl & Risak, 2016).

- This optimization leads to a more efficient distribution of work, though it also raises concerns about the transparency of the algorithmic decision-making process.

2. **Personalization of Job Opportunities**:
   - AI-driven platforms personalize job recommendations based on workers' past experiences, preferences, and performance (Harris & Krueger, 2015).
   - This personalization enhances worker engagement but may also lead to a narrowing of job exposure, potentially limiting diverse work opportunities.

3. **Dynamic Pricing Models**:
   - Many gig platforms use AI to implement dynamic pricing strategies, adjusting prices in real-time based on demand, supply, and other market factors (Chen et al., 2016).
   - While this can maximize earnings and efficiency, it can also lead to unpredictable income for gig workers.

4. **Quality Control and Performance Monitoring**:
   - AI tools are used to monitor the quality of work performed by gig workers, providing feedback and ratings that can impact future job opportunities (De Stefano, 2016).
   - This automated oversight ensures consistent service quality but may also exert pressure on workers to conform to machine-driven assessments of performance.

5. **Facilitating Remote and Flexible Work**:
   - AI-driven platforms have been instrumental in facilitating remote and flexible work arrangements, especially significant during the COVID-19 pandemic (Kässi & Lehdonvirta, 2018).
   - This shift has helped sustain the gig economy during challenging times but also highlights the need for digital connectivity and skills.

6. **Data Privacy and Security Concerns**:

- The extensive data collection by AI-driven platforms raises significant concerns regarding privacy and data security for gig workers (Cohen & Sundararajan, 2015).
- Ensuring data protection and complying with regulations like GDPR is crucial for these platforms.

7. Future of Work and Automation:

- AI-driven platforms are not only intermediaries but also innovators, potentially automating certain types of gig work in the future (Sundararajan, 2016).
- This prospect raises questions about the long-term sustainability of gig work and the need for workers to adapt to evolving technology.

II. Impact on Worker Autonomy and Flexibility:

The integration of Artificial Intelligence (AI) in the gig economy has profound implications for worker autonomy and flexibility. AI-driven platforms are reshaping how gig work is structured, with both positive and negative effects on the workforce.

1. Enhanced Flexibility in Work Scheduling:

- AI algorithms allow gig workers to have more control over their work schedules, facilitating on-demand work availability (De Stefano, 2016).
- This flexibility is particularly beneficial for those seeking to balance work with other commitments, although it may come at the cost of job security and benefits (Johnston & Land-Kazlauskas, 2018).

2. Autonomy in Task Selection:

- Platforms powered by AI provide workers with the ability to choose tasks that match their skills and interests (Wood et al., 2019).
- However, the autonomy in task selection is often limited by the algorithm's control over which tasks are visible to

which workers, potentially leading to biased job allocations (Prassl, 2018).

3. **Customized Job Recommendations**:
   - AI algorithms can offer personalized job recommendations, improving the match between a worker's skills and available gigs (Horton & Zeckhauser, 2016).
   - This customization enhances worker satisfaction but may also narrow the variety of work opportunities, pigeonholing workers into specific types of tasks.

4. **Erosion of Traditional Employment Protections**:
   - The gig economy, facilitated by AI, often operates outside traditional employment structures, lacking protections like minimum wage, overtime, and unemployment insurance (Harris & Krueger, 2015).
   - This shift challenges the balance between autonomy and the security traditionally provided by employment.

5. **Data-Driven Performance Management**:
   - AI systems monitor and evaluate worker performance, with ratings affecting future work opportunities (Lee et al., 2015).
   - While this can promote high-quality work, it may also lead to a sense of continuous surveillance, impacting worker autonomy.

6. **Influence on Work-Life Balance**:
   - The gig economy's flexibility allows workers to better manage work-life balance, but it also blurs the boundaries between personal and professional life (Duggan et al., 2020).
   - Workers may find themselves always "on call," potentially leading to increased stress and burnout.

7. **Skill Development Opportunities**:

- ○ AI-driven gig platforms can offer avenues for skill development and career progression (Kässi & Lehdonvirta, 2018).
- ○ However, the responsibility for professional growth largely rests on the individual, requiring proactive engagement in continuous learning.

III. **AI and Workforce Optimization**:

The advent of Artificial Intelligence (AI) in the gig economy significantly influences workforce optimization. This application of AI is reshaping the dynamics of how gig work is allocated, managed, and evaluated.

1. **Efficient Resource Allocation**:
   - ○ AI algorithms optimize the matching of gig tasks with suitable workers based on skills, availability, and past performance (Aloisi & De Stefano, 2021).
   - ○ This efficient allocation can lead to higher productivity but may also reduce the diversity of work experiences for gig workers (Wood et al., 2019).
2. **Dynamic Pricing Mechanisms**:
   - ○ AI enables dynamic pricing strategies, adjusting pay rates based on demand, worker availability, and urgency (Lee et al., 2015).
   - ○ While this can maximize earnings during peak times, it can also lead to unpredictability in income for workers.
3. **Predictive Analytics in Hiring**:
   - ○ Predictive analytics are used to forecast demand and determine the number of gig workers needed, improving operational efficiency (Horton & Zeckhauser, 2016).
   - ○ However, reliance on predictive models can lead to over- or under-hiring, affecting job stability for workers.
4. **Automated Performance Monitoring**:

- AI systems continually monitor worker performance, providing feedback and identifying areas for improvement (Prassl, 2018).
- This constant monitoring can enhance quality but may also contribute to a sense of surveillance and pressure.

5. **Real-Time Feedback and Adaptation**:
- Real-time feedback mechanisms enabled by AI allow for quick adaptation to changing work conditions and client needs (De Stefano, 2016).
- This agility benefits the service quality but requires workers to be continuously adaptable and flexible.

6. **Enhancing Worker Engagement and Retention**:
- AI can analyze worker preferences and job satisfaction, helping platforms to tailor the work experience to increase engagement and retention (Kässi & Lehdonvirta, 2018).
- The challenge lies in balancing algorithm-driven work assignments with worker autonomy and preference.

7. **Risk of Algorithmic Bias**:
- AI algorithms, if not properly designed and monitored, can perpetuate biases in workforce optimization, unfairly favoring or disadvantaging certain groups (Wood et al., 2019).
- Ensuring fairness and transparency in AI systems is crucial to maintain trust and equity in the gig economy.

IV. The Future of Work and Employment:

The integration of AI into the gig economy is redefining the future landscape of work and employment. This transformation brings both opportunities and challenges that will shape labor markets in profound ways.

1. **Increased Job Flexibility and Independence**:

- AI-driven platforms in the gig economy offer greater flexibility, enabling individuals to choose when, where, and how much they work (De Stefano, 2016).
- This flexibility, however, may lead to inconsistent income and lack of traditional employment benefits (Wood et al., 2019).

2. **Changing Nature of Skills and Training**:

- The gig economy, fueled by AI, demands a shift in skill sets, emphasizing adaptability, digital literacy, and continuous learning (Brynjolfsson & McAfee, 2014).
- Workers need to be upskilled or reskilled to remain competitive, which raises questions about who bears the responsibility for this training (Kässi & Lehdonvirta, 2018).

3. **Decentralization of Workforces**:

- AI enables work to be distributed globally, leading to a decentralized workforce (Manyika et al., 2016).
- This globalization of gig work presents opportunities for workers worldwide but also intensifies competition and may drive down wages (Graham et al., 2017).

4. **Economic Implications and Job Creation**:

- AI in the gig economy has the potential to create new job categories, fostering economic growth (Autor, 2015).
- However, there is also the risk of job displacement, especially for routine and manual tasks susceptible to automation (Acemoglu & Restrepo, 2020).

5. **Social and Legal Implications**:

- The rise of gig work challenges traditional employment models, leading to debates over worker classification, rights, and protections (Prassl, 2018).
- Governments and policymakers are called upon to update labor laws and social security systems to adapt to these new forms of work (Harris & Krueger, 2015).

6. **Ethical and Fairness Considerations**:

- AI algorithms need to be designed ethically to ensure fairness in job allocation and compensation (Lee et al., 2015).
- Addressing algorithmic bias and ensuring transparency becomes crucial to maintain trust and equity in the gig economy (Wood et al., 2019).

7. **Impact on Work-Life Balance**:
- While the gig economy offers flexibility, it also blurs the line between personal and professional life, potentially leading to overwork and stress (De Stefano, 2016).
- Achieving a healthy work-life balance remains a challenge for gig workers operating in an AI-driven environment.

V. **Regulatory and Ethical Considerations**:

The merging of AI with the gig economy has significant regulatory and ethical implications, necessitating a careful examination of how these technologies are integrated into the workforce. This intersection poses unique challenges that require a rethinking of traditional regulatory frameworks and ethical norms.

1. **Worker Classification and Rights**:
- The gig economy's employment models challenge the traditional employer-employee relationship, raising questions about worker classification (Prassl & Risak, 2016).
- Regulations need to address whether gig workers should be classified as employees or independent contractors, impacting their rights and benefits (De Stefano, 2016).

2. **Data Privacy and Protection**:
- AI systems in the gig economy rely heavily on data, raising concerns about privacy and data protection (Pasquale, 2015).
- Regulations such as the General Data Protection Regulation (GDPR) in the EU provide a framework, but specific

guidelines for gig work are needed (Graef, Husovec, & Purtova, 2018).

3. **Algorithmic Transparency and Accountability**:

   ○ The use of AI algorithms in gig platforms necessitates transparency to ensure fairness in job assignments and wage determinations (Lee et al., 2015).

   ○ Regulatory frameworks must ensure that these algorithms do not reinforce biases or discriminatory practices (Barocas, Hardt, & Narayanan, 2019).

4. **Ethical Use of AI**:

   ○ There's a need for ethical guidelines to govern the use of AI in gig platforms, focusing on fairness, non-discrimination, and respect for worker autonomy (Martin, 2019).

   ○ Ethical considerations include ensuring equitable access to gig work and protecting workers from exploitation (Wood et al., 2019).

5. **Health and Safety Regulations**:

   ○ Gig workers often lack the health and safety protections afforded to traditional employees (De Stefano, 2016).

   ○ Regulatory bodies need to adapt existing health and safety laws to the realities of gig work, ensuring adequate protection for all workers.

6. **Social Security and Benefits**:

   ○ The transient nature of gig work raises questions about social security and long-term benefits, including pensions, healthcare, and unemployment insurance (Harris & Krueger, 2015).

   ○ Policymakers face the challenge of redesigning social security systems to accommodate the non-standard employment models of the gig economy (Sundararajan, 2016).

7. **International Regulatory Considerations**:

- The global nature of the gig economy requires international cooperation in regulatory approaches to address cross-border challenges (Graham et al., 2017).
- Harmonizing regulations across countries is crucial to ensure consistent protections for gig workers regardless of location (Sundararajan, 2016).

VI. **Economic Implications and Inequality**:

The integration of Artificial Intelligence (AI) into the gig economy has significant economic implications, particularly concerning inequality. This fusion is altering traditional employment structures and economic models, leading to both opportunities and challenges.

1. **Income Disparity**:
   - The gig economy, augmented by AI, can exacerbate income disparities. High-skilled workers may have access to more lucrative opportunities, while low-skilled workers face a race to the bottom in terms of wages (Graham, Hjorth, & Lehdonvirta, 2017).
   - This disparity is heightened by the digital divide, where access to technology and digital literacy determines earning potential (Van Dijk, 2020).
2. **Job Displacement and Creation**:
   - AI in the gig economy may lead to job displacement, especially in roles susceptible to automation (Frey & Osborne, 2017).
   - However, it also creates new job opportunities, particularly in tech-driven sectors and roles that require human creativity and empathy (Brynjolfsson & McAfee, 2014).
3. **Market Concentration and Monopsony**:
   - The gig economy, powered by AI, tends to create market concentrations where a few large platforms dominate, potentially leading to monopsony-like conditions,

influencing wages and working conditions (Kenney & Zysman, 2016).

- This concentration can limit workers' bargaining power and contribute to inequality (Srnicek, 2017).

4. **Worker Exploitation Risks**:

- The lack of regulation in the gig economy can lead to the exploitation of workers, with AI exacerbating this through opaque algorithmic management (Wood et al., 2019).
- Issues such as unfair performance evaluations, wage theft, and lack of benefits are prevalent (Rosenblat & Stark, 2016).

5. **Economic Instability for Workers**:

- Gig workers often face economic instability due to the precarious nature of gig work, exacerbated by AI-driven demand fluctuations (De Stefano, 2016).
- This instability impacts their ability to plan financially for the long term, contributing to economic inequality (Kalleberg & Dunn, 2016).

6. **Access to Benefits and Social Safety Nets**:

- Traditional employment benefits such as health insurance, retirement plans, and unemployment benefits are often not available to gig workers (Harris & Krueger, 2015).
- This lack of access contributes to widening economic inequality, as gig workers are left more vulnerable compared to traditional employees (Sundararajan, 2016).

7. **Global Inequality**:

- The global nature of the gig economy means that workers from lower-income countries might face unfair competition and lower wages compared to those in higher-income countries (Graham et al., 2017).
- This global disparity can lead to a form of digital colonialism, where wealth is extracted from less economically developed regions (Qiu, Gregg, & Crawford, 2014).

VII. **The Role of AI in Skill Development and Training**:

The role of Artificial Intelligence (AI) in skill development and training within the gig economy represents a transformative shift in how workers acquire and enhance their capabilities. AI technologies offer unique opportunities for personalized learning, skill adaptation, and career progression in the gig economy.

1. **Personalized Learning Paths**:
   - AI enables the creation of tailored training programs that adapt to an individual's learning pace, style, and existing skill set (Zhou et al., 2020).
   - Such personalization ensures that gig workers can efficiently acquire skills relevant to their career aspirations and market demands (Hwang, 2014).

2. **Real-Time Skill Gap Analysis**:
   - AI systems can analyze job market trends and identify emerging skills gaps, guiding gig workers towards acquiring in-demand skills (Bessen, 2019).
   - This real-time analysis aids workers in staying competitive and relevant in the rapidly evolving gig economy (Kaplan & Haenlein, 2019).

3. **Adaptive Learning Technologies**:
   - AI-driven adaptive learning technologies can adjust content and teaching methodologies based on the learner's performance and feedback (Xie et al., 2019).
   - This adaptability enhances the learning experience, making it more effective for diverse learners with varying backgrounds (Hwang & Lai, 2017).

4. **Continuous Learning and Upskilling**:
   - AI facilitates continuous learning and upskilling opportunities, essential in the gig economy where job roles and required skills are constantly changing (Frey & Osborne, 2017).

- Platforms leveraging AI can offer ongoing training modules to help workers adapt to new technologies and methodologies (Manyika et al., 2017).

5. **Enhanced Accessibility and Inclusivity**:

- AI-driven platforms can make learning resources more accessible to a broader range of individuals, including those with disabilities or from remote locations (Burgsteiner et al., 2016).
- This inclusivity is crucial in ensuring equitable skill development opportunities within the gig economy (Zhao et al., 2019).

6. **Career Pathing and Progression**:

- AI can assist gig workers in career pathing by analyzing job trajectories and market trends to provide recommendations on potential career advancements (Brynjolfsson & McAfee, 2014).
- This guidance is particularly valuable in the gig economy, where traditional career progression paths may be less clear (De Stefano, 2016).

7. **Cost-Effective and Scalable Training Solutions**:

- AI-driven training solutions are often more cost-effective and scalable compared to traditional training methods, making them ideal for gig economy workers who may not have access to corporate training programs (Bughin et al., 2018).
- This scalability ensures a wide-reaching impact, benefiting a large number of gig workers (Ferrari et al., 2018).

Rise of freelancing, remote work, and platform-based employment:

The intersection of the gig economy and AI has significantly influenced the rise of freelancing, remote work, and platform-based employment. This transformation is reshaping the global labor market, offering new opportunities and challenges.

1. Growth of Freelancing and Independent Work:

The emergence of AI in the gig economy has played a pivotal role in the growth of freelancing and independent work. This transformation reflects a significant shift in the labor market dynamics, presenting new opportunities and challenges.

1. **Facilitation of Freelance Work through AI:**
   - AI-driven platforms have revolutionized the way freelancers connect with potential clients, offering advanced job-matching algorithms and personalized project recommendations (De Stefano, 2016).
   - These platforms utilize AI to analyze market trends and skill demands, enabling freelancers to adapt and develop relevant skills (Katz & Krueger, 2019).
2. **Increased Accessibility and Global Reach:**
   - AI technologies have made it easier for freelancers to access global markets, breaking down geographical and logistical barriers (Smith & Leberstein, 2015).
   - This global reach has expanded the potential client base for freelancers, allowing them to work with a diverse range of clients from different regions and industries (Howell, 2016).
3. **Enhancement of Freelance Work Efficiency:**
   - AI tools assist in automating routine tasks such as billing, scheduling, and client communication, thus increasing the efficiency and productivity of freelancers (Bloom, 2017).
   - These technologies also provide data-driven insights, helping freelancers to optimize their pricing strategies and project management (Prassl & Risak, 2016).

4. **Impact on Work-Life Balance**:
   - The flexibility offered by freelancing, empowered by AI, has led to improved work-life balance for many independent workers (Johnston & Land-Kazlauskas, 2018).
   - However, this flexibility can also blur the lines between personal and professional life, posing challenges in terms of time management and work stress (Eurofound, 2020).

5. **Skill Development and Adaptability**:
   - AI has necessitated continuous learning and adaptability among freelancers, as they need to stay abreast of technological advancements and changing market needs (Harris & Krueger, 2015).
   - Various online platforms offer AI-driven skill development courses, making it easier for freelancers to acquire new skills and enhance their marketability (Aloisi, 2015).

II. Expansion of Remote Work:

The rise of AI in the gig economy has significantly contributed to the expansion of remote work. This shift represents a fundamental change in traditional work environments and practices, reflecting the evolving nature of the workforce.

1. **Enabling Remote Work Through Advanced Technologies**:
   - AI and machine learning technologies have facilitated remote work by enhancing communication tools and project management software, making it easier for teams to collaborate effectively from various locations (Bloom, 2017).
   - Virtual assistance and automation tools, powered by AI, help in managing schedules, emails, and tasks, thus streamlining the workflow for remote workers (Ferrari, 2018).

2. **Increased Productivity and Efficiency**:
   - Studies have shown that remote work can lead to increased productivity, as workers are less exposed to office

distractions and can create a personalized work environment (Ozimek, 2020).

- AI-driven analytics tools provide insights on work patterns and productivity, enabling remote workers to optimize their schedules for better efficiency (Eurofound, 2020).

3. **Expansion of Talent Pool and Diversity**:

- Remote work allows companies to access a global talent pool, not limited by geographical constraints, leading to more diverse and skilled workforces (Kaplan et al., 2020).
- This diversity enhances creativity and innovation within teams, as employees bring varied perspectives and experiences to the table (Erickson, 2019).

4. **Challenges in Remote Work**:

- Despite the advantages, remote work also presents challenges, including potential isolation, difficulty in separating work from personal life, and reliance on digital communication (Mann & Holdsworth, 2020).
- Ensuring cybersecurity and data privacy becomes paramount with remote work, as employees access company resources from different networks (Kowalski & Swanson, 2017).

5. **Work-Life Balance and Employee Well-being**:

- Remote work offers flexibility, contributing to a better work-life balance, but it also requires self-discipline and effective time management (Golden, 2018).
- Companies are increasingly using AI-driven tools for monitoring employee well-being and offering support for mental health and stress management (Parker, 2019).

III. Platform-Based Employment Ecosystems:

The integration of AI into the gig economy has catalyzed the growth of platform-based employment ecosystems, fundamentally altering the landscape of freelance and contract work.

1. **Development of Digital Platforms**:
   - AI-driven platforms have revolutionized the way freelancers connect with potential clients, offering efficient job matching, skill assessment, and project management tools (De Stefano, 2020).
   - These platforms utilize sophisticated algorithms to personalize job recommendations based on a freelancer's skills, experience, and preferences, enhancing the job search process (Howcroft & Bergvall-Kåreborn, 2019).

2. **Enhanced Market Efficiency**:
   - AI technologies in these platforms streamline administrative tasks like billing and contracts, allowing freelancers to focus more on their work (Johnston & Land-Kazlauskas, 2018).
   - The platforms provide a level of security for both freelancers and clients through reputation systems and dispute resolution mechanisms, fostering a more trustworthy environment (Burtch et al., 2018).

3. **Data-Driven Insights for Skill Development**:
   - AI algorithms analyze market trends and provide freelancers with insights on in-demand skills, guiding them in their professional development (Fleming, 2020).
   - This adaptive learning and development approach helps workers stay competitive in an evolving job market (Spencer et al., 2020).

4. **Economic Impact and Worker Classification**:
   - While these platforms contribute to economic growth by facilitating gig work, they also raise questions about worker classification and benefits (Katz & Krueger, 2019).
   - The debate continues over whether gig workers should be classified as independent contractors or employees, impacting their access to benefits and protections (Sundararajan, 2017).

5. **Diverse Opportunities and Challenges**:
   - Platform-based employment offers opportunities for workers from varied backgrounds to access global job markets, potentially reducing barriers to entry (Prassl & Risak, 2018).
   - However, it also introduces challenges related to job stability, income predictability, and the digital divide, as not all workers have equal access to these platforms (Graham et al., 2017).

IV. **AI in Enhancing Platform Efficiency**:

The rise of freelancing, remote work, and platform-based employment has been significantly influenced by the integration of Artificial Intelligence (AI), particularly in enhancing the efficiency of these platforms.

1. **Automated Job Matching**:
   - AI algorithms have become essential in matching freelancers with suitable jobs. These systems analyze vast amounts of data, including job descriptions and freelancer profiles, to recommend the most relevant opportunities (Deng et al., 2020).
   - This automation reduces the time spent on job searching, leading to increased productivity and job satisfaction among freelancers (Burtch et al., 2018).

2. **Efficient Resource Allocation**:
   - AI-driven platforms optimize resource allocation, ensuring that tasks are assigned to freelancers with the appropriate skill sets (Howcroft & Bergvall-Kåreborn, 2019).
   - This enhances the quality of work delivered and minimizes the time and costs associated with mismatches in the job allocation process (Prassl & Risak, 2018).

3. **Real-Time Data Analysis and Forecasting**:

- AI tools analyze market trends and predict future work opportunities, enabling freelancers to adapt to changing market demands (Sundararajan, 2017).
- These insights help freelancers to strategically develop skills that are likely to be in demand, thereby maintaining their competitiveness in the market (Spencer et al., 2020).

4. **Improved User Experience**:

- AI contributes to a more personalized user experience on gig platforms. It helps in tailoring the platform's interface and functionalities to individual user preferences, enhancing usability and engagement (Katz & Krueger, 2019).
- This personalization extends to customer support, where AI-driven chatbots provide immediate assistance, improving the overall user experience (Johnston & Land-Kazlauskas, 2018).

5. **Scalability and Flexibility**:

- The use of AI enables gig platforms to scale efficiently, accommodating the growing number of freelancers and the expanding variety of job types (Graham et al., 2017).
- AI systems provide the flexibility to adapt to different languages and regional job markets, making the platforms globally accessible (De Stefano, 2020).

V. **Challenges and Regulatory Implications**:

The intersection of freelancing, remote work, and platform-based employment with Artificial Intelligence (AI) presents unique challenges and regulatory implications.

1. **Worker Classification and Legal Rights**:

- The gig economy blurs traditional boundaries between employees and independent contractors. This ambiguity raises questions about legal rights, benefits, and protections for gig workers (De Stefano, 2020).

- Regulatory frameworks struggle to categorize these workers, impacting their access to labor protections, social security, and health benefits (Prassl & Risak, 2018).

2. **Data Privacy and Security**:

- The extensive use of personal data in AI-driven gig platforms raises significant privacy concerns. There's a need for regulations to safeguard worker data against misuse and breaches (Möhlmann & Zalmanson, 2017).
- This includes addressing issues related to data ownership, consent, and transparency in data handling (Barrett et al., 2019).

3. **Algorithmic Bias and Fairness**:

- AI algorithms can inadvertently perpetuate biases, leading to unfair job allocations or discriminatory practices (Katz & Krueger, 2019).
- Regulations are needed to ensure algorithmic transparency and fairness, preventing biases based on race, gender, or other protected characteristics (Sundararajan, 2017).

4. **Income Stability and Job Security**:

- The gig economy is often characterized by inconsistent income and lack of job security. Policies must address these issues to ensure financial stability for gig workers (Howcroft & Bergvall-Kåreborn, 2019).
- This includes consideration of minimum wage standards, unemployment benefits, and income stabilization mechanisms (Graham et al., 2017).

5. **Health and Safety Regulations**:

- Traditional health and safety regulations are often not applicable in the decentralized gig economy. New frameworks are needed to protect workers in diverse and remote working environments (Burtch et al., 2018).

- This is particularly crucial in contexts where AI-driven platforms may encourage longer working hours or higher workloads (Johnston & Land-Kazlauskas, 2018).

6. **Intellectual Property Rights**:

- Issues around intellectual property rights in gig work, especially in creative and technical fields, necessitate clear regulatory guidelines to protect the interests of both freelancers and clients (Deng et al., 2020).

How AI is reshaping the dynamics of flexible work:

Artificial Intelligence (AI) is significantly reshaping the dynamics of flexible work, especially within the gig economy. This transformation is characterized by several key aspects:

1. **Enhanced Job Matching and Allocation**:

The integration of Artificial Intelligence (AI) into the gig economy has notably revolutionized the way job matching and allocation are conducted. This transformation can be understood through several key dimensions:

1. **Sophisticated Algorithms for Precise Matching**:

- AI employs complex algorithms to analyze vast amounts of data, including worker skills, job requirements, and historical performance, to create highly accurate matches (Deng et al., 2020). This precise matching ensures that gig workers are aligned with jobs that suit their expertise and interests, leading to better job satisfaction and performance.
- For example, platforms like LinkedIn use AI to suggest job opportunities that align closely with the user's profile and past job searches (Barrett et al., 2019).

2. **Efficient Resource Allocation**:
   - AI enables platforms to optimize resource allocation by predicting demand and assigning workers accordingly. This not only maximizes the efficiency of the workforce but also ensures that there is a balance between supply and demand, reducing idle times for gig workers (Katz & Krueger, 2019).
   - In the context of ride-sharing services like Uber, AI algorithms predict demand patterns and guide drivers to areas where they are most likely to find passengers (Prassl & Risak, 2018).

3. **Personalization and Worker Preferences**:
   - AI systems can take into account personal preferences and constraints of workers, such as desired work hours, location preferences, and types of tasks. This personalization leads to a more worker-centric approach, enhancing the gig experience (Möhlmann & Zalmanson, 2017).
   - Platforms like TaskRabbit use AI to match freelancers with tasks that not only fit their skillset but also their personal scheduling preferences (Stanton & Thomas, 2018).

4. **Real-time Matching and Dynamic Adjustments**:
   - The real-time capabilities of AI ensure that job matching is a continuous and dynamic process. AI algorithms can adjust recommendations based on changing circumstances, such as sudden availability or cancellation of tasks (Sundararajan, 2017).
   - This dynamic matching is particularly beneficial in fast-paced gig sectors, like delivery services, where real-time adjustments are crucial for operational efficiency.

5. **Feedback Loops for Continuous Improvement**:
   - AI systems incorporate feedback mechanisms, allowing them to learn and improve over time. Positive and negative feedback from both workers and employers refine the job

matching algorithms, leading to more accurate matches in the future (Howcroft & Bergvall-Kåreborn, 2019).

6. **Reducing Bias in Job Allocation**:
   - AI has the potential to reduce human biases in job allocation. By focusing on data-driven criteria, AI can help ensure that job assignments are based on merit and suitability rather than unconscious biases (Graham et al., 2017).

II. Flexible Scheduling and Workload Management:

Artificial Intelligence (AI) is significantly impacting the gig economy, particularly in the areas of flexible scheduling and workload management. This shift is characterized by several critical aspects:

1. **AI-Enabled Dynamic Scheduling**:
   - AI systems allow for dynamic scheduling, accommodating the fluctuating availability of gig workers and the varying demand for services (Prassl, 2018). This flexibility benefits workers who seek control over their work hours and life balance.
   - For example, platforms like Upwork and Fiverr use AI to suggest projects and gigs that fit within the available time slots of freelancers, allowing them to manage their schedules more effectively (De Stefano, 2016).
2. **Predictive Workload Management**:
   - AI's predictive capabilities enable platforms to forecast workloads and help workers plan their schedules in advance (Howcroft & Bergvall-Kåreborn, 2019). This foresight is crucial in managing periods of high and low demand, ensuring a steady flow of work for gig workers.
   - Ride-sharing apps like Lyft use AI to predict rider demand, helping drivers plan their working hours around predicted busy times (Katz & Krueger, 2019).
3. **Customization of Work Preferences**:

- ○ AI systems cater to individual worker preferences, including preferred types of tasks, working hours, and workload intensity. This customization enhances job satisfaction and work-life balance (Möhlmann & Zalmanson, 2017).
- ○ For instance, freelance platforms can use AI to align gig opportunities with the personal preferences and past work patterns of their users (Deng et al., 2020).

4. **AI-Assisted Workload Distribution**:

- ○ AI algorithms can distribute workload among gig workers in a more balanced and equitable manner. This distribution helps in avoiding overburdening some workers while underutilizing others (Stanton & Thomas, 2018).
- ○ Task allocation in platforms like TaskRabbit can be optimized using AI to ensure a fair distribution of work among available freelancers.

5. **Real-Time Adjustment to Workload Changes**:

- ○ AI systems can quickly adapt to real-time changes in workload, such as last-minute cancellations or additional task requests. This agility helps gig workers to adjust their schedules and workload on the fly (Sundararajan, 2017).
- ○ In delivery services, for example, AI can immediately reallocate tasks if a worker becomes unavailable, ensuring timely deliveries and efficient use of the workforce.

6. **Reducing the Risk of Overwork and Burnout**:

- ○ By intelligently managing workloads and schedules, AI helps mitigate the risk of overwork and burnout, which is a common concern in the gig economy (Graham et al., 2017).
- ○ AI can monitor work hours and suggest breaks or time off when patterns indicate potential burnout.

III. Skill Development and Career Progression:

The intersection of AI and the gig economy is reshaping the landscape of skill development and career progression in several key ways:

1. **AI-Driven Personalized Learning Paths**:
   - AI enables the creation of personalized learning paths for gig workers, facilitating skill development based on individual needs and career goals (Bessen, 2019). This approach helps workers continuously adapt and upskill in a rapidly evolving job market.
   - Platforms like Coursera and Udemy utilize AI to recommend courses and training programs tailored to the skill gaps and interests of freelancers, thus aiding in their professional growth (Frey & Osborne, 2017).

2. **Automated Skill Assessment and Gap Analysis**:
   - AI tools can efficiently assess a worker's current skill set and identify areas for improvement, allowing for targeted skill development (Kaplan & Haenlein, 2019). This analysis helps workers stay competitive and meet the evolving demands of the gig economy.
   - For example, LinkedIn's AI algorithms suggest skills that are in demand for particular roles, helping gig workers understand where they need to upskill (Autor, 2015).

3. **Facilitation of Micro-Credentials and Certifications**:
   - AI enables the integration of micro-credentials and certifications into gig platforms, providing workers with tangible evidence of their skills and learning achievements (Oliver, 2019). These credentials are increasingly recognized by employers and can significantly enhance career progression.
   - Platforms like edX offer AI-driven recommendations for micro-credential programs, which gig workers can use to showcase their expertise in specific areas (Manyika et al., 2017).

4. **Career Path Forecasting**:
   - AI can predict future industry trends and suggest potential career paths for gig workers, helping them to make informed decisions about their professional development (Brynjolfsson & Mitchell, 2017).
   - Tools like IBM's Watson offer insights into emerging job roles and the skills required for them, guiding gig workers in planning their career trajectories (Susskind & Susskind, 2015).

5. **Enhanced Visibility and Recognition**:
   - AI algorithms in gig platforms can highlight workers' skills and accomplishments to potential clients, enhancing their visibility and opportunities for career advancement (Burtch et al., 2018).
   - Platforms like Behance use AI to showcase the portfolios of creative professionals, making it easier for them to get noticed by potential clients and collaborators.

6. **Continuous Feedback and Performance Analysis**:
   - AI systems provide continuous feedback and performance analysis, helping gig workers identify strengths and areas for improvement. This ongoing assessment is crucial for career growth in a competitive market (Agrawal et al., 2018).
   - Feedback mechanisms in platforms like Upwork can be AI-enhanced to give more detailed insights into a freelancer's performance and areas for development.

IV. **Performance Analysis and Feedback**:

The integration of AI in the gig economy is revolutionizing performance analysis and feedback, reshaping the dynamics of flexible work in significant ways:

1. **Real-Time Performance Tracking**:

- AI systems facilitate real-time tracking of gig workers' performance, enabling immediate feedback and adjustments. This feature is particularly useful in dynamic work environments where timely responses are crucial (De Stefano, 2018).
- Platforms like TaskRabbit use AI algorithms to monitor task completion rates and quality, providing workers with instant feedback that helps them improve their services (Deng & Joshi, 2016).

2. **Data-Driven Feedback**:

- AI enables the collection and analysis of vast amounts of performance data, leading to more objective and comprehensive feedback for gig workers (Autor, 2015). This data-driven approach helps in identifying specific areas for improvement.
- Uber's rating system, powered by AI, analyzes patterns in driver performance, offering them constructive feedback to enhance their service quality (Lee et al., 2015).

3. **Personalized Improvement Suggestions**:

- AI can provide personalized suggestions for improvement based on individual performance metrics, work habits, and past experiences (Kaplan & Haenlein, 2019). This customization allows workers to focus on areas most relevant to their skill set and career goals.
- Freelancer platforms like Upwork employ AI to suggest specific skill development courses or assignments that align with a freelancer's performance history (Manyika et al., 2017).

4. **Predictive Performance Modeling**:

- AI can predict future performance trends based on historical data, helping gig workers anticipate challenges and prepare accordingly (Bessen, 2019). This predictive capability is key in maintaining a competitive edge in the gig economy.

- Predictive models in platforms like Airbnb anticipate guest needs and preferences, advising hosts on how to tailor their services for better reviews and improved performance (Gimpel et al., 2018).

5. **Enhanced Quality Assurance**:

- AI tools assist in maintaining high standards of quality assurance by consistently monitoring work output. This ensures a uniform quality of service across the platform, benefiting both gig workers and clients (Agrawal et al., 2018).
- Quality control algorithms on platforms like Etsy help sellers understand customer satisfaction levels and improve their product listings based on AI-generated insights (Frenken & Schor, 2017).

6. **Automated Conflict Resolution**:

- AI can assist in resolving disputes by analyzing performance data and communication records, leading to fair and efficient conflict resolution (Susskind & Susskind, 2015). This automation reduces the time and resources spent on resolving workplace issues.
- Conflict resolution systems in platforms like eBay use AI to mediate between buyers and sellers, often resolving issues without human intervention (Malhotra & Van Alstyne, 2014

V. **Predictive Analytics for Demand Forecasting**:

The role of AI in reshaping the dynamics of flexible work is profoundly evident in the realm of predictive analytics for demand forecasting. This technology is transforming how gig economy platforms and workers anticipate and respond to market needs:

1. **Demand Prediction**:
   - AI-driven predictive analytics allow gig economy platforms to forecast demand trends accurately, enhancing their ability to manage workforce supply (Manyika et al., 2017). For example, ride-sharing apps like Uber use AI to predict rider demand, ensuring an optimal number of drivers in various locations (Cramer & Krueger, 2016).

2. **Resource Allocation Optimization**:
   - By predicting demand, AI aids in the efficient allocation of resources. This optimization ensures that gig workers are directed towards areas with the highest demand, maximizing their earnings potential (Hall & Krueger, 2018).
   - Platforms like DoorDash employ AI to forecast order volumes, helping them assign deliveries to drivers in the most efficient manner (Chen et al., 2019).

3. **Anticipating Consumer Behavior**:
   - AI algorithms analyze historical data to anticipate consumer behavior patterns. This insight allows gig workers and platforms to adapt their services to meet evolving customer expectations (Agrawal et al., 2018).
   - Airbnb uses AI to analyze travel trends, advising hosts on how to adjust their offerings based on predicted traveler preferences (Guttentag, 2015).

4. **Seasonal and Event-Based Predictions**:
   - AI systems excel at identifying patterns related to seasonal changes or special events, enabling gig platforms to prepare for fluctuations in demand (Kaplan & Haenlein, 2019). This foresight is critical for maintaining service quality during peak times.
   - Event prediction models in apps like Lyft anticipate increased demand during concerts or sports events, advising drivers on where to position themselves (Greenwood & Wattal, 2017).

5. **Dynamic Pricing Strategies**:
   - Predictive analytics play a crucial role in implementing dynamic pricing strategies, which adjust prices in real-time based on demand. This approach helps gig economy platforms balance supply and demand effectively (Chen & Sheldon, 2016).
   - Uber's surge pricing algorithm, based on predictive demand models, adjusts fares to match driver availability with rider demand (Lam & Liu, 2018).

6. **Enhancing Worker Preparedness**:
   - Gig workers benefit from AI's demand forecasting by gaining insights into potential work opportunities. This preparedness helps them plan their schedules and locations to maximize income (De Stefano, 2018).
   - TaskRabbit's AI system provides its freelancers with information about the most in-demand tasks, enabling them to tailor their services accordingly (Huws et al., 2017).

VI. Enhanced Customer Experience:

The integration of AI into the gig economy has significantly impacted the dynamics of flexible work, particularly in enhancing customer experience. This transformation is evident in various aspects:

1. **Personalized Service Offerings**:
   - AI enables gig platforms to offer highly personalized services to customers. By analyzing customer data, AI can suggest services tailored to individual preferences, enhancing customer satisfaction (Huang & Rust, 2018). For instance, platforms like TaskRabbit can suggest specific freelancers to users based on their past hiring history and preferences (Zhu & Liu, 2018).

2. **Improved Service Quality and Responsiveness**:

- AI-driven chatbots and customer service tools have revolutionized customer support in the gig economy. These tools provide quick and efficient responses to customer inquiries, improving the overall service quality (Van Doorn et al., 2017). For example, Airbnb uses AI to assist guests and hosts in resolving issues, thereby enhancing the user experience (Guttentag, 2015).

3. **Enhanced Matching Algorithms**:

   - AI algorithms are crucial for effectively matching customers with the most suitable gig workers or services. This precise matching ensures that customers receive the service that best fits their needs, leading to higher satisfaction levels (Agrawal et al., 2018). Uber's algorithm, for instance, matches riders with drivers to minimize wait times and improve ride experiences (Cramer & Krueger, 2016).

4. **Real-Time Feedback and Improvement**:

   - AI systems allow for the collection and analysis of real-time feedback from customers, which gig platforms can use to continuously improve their services (Bughin et al., 2017). Platforms like Upwork use customer feedback to refine their freelancer recommendations, ensuring better matches for future projects (De Stefano, 2018).

5. **Predictive Customer Support**:

   - By analyzing customer behavior and interaction patterns, AI can predict potential issues and provide proactive solutions, further enhancing the customer experience (Kaplan & Haenlein, 2019). For example, delivery services like Postmates use AI to anticipate and address delivery issues before they escalate (Chen et al., 2019).

6. **Seamless Integration Across Services**:

   - AI facilitates the integration of various services, providing a seamless experience for customers. This integration can range from transportation to food delivery, all accessible through a single platform (Manyika et al., 2017). An example is how Uber integrates ride services with Uber Eats, offering a cohesive experience to users (Hall & Krueger, 2018).

VII. Safety and Compliance Monitoring:

Artificial Intelligence (AI) is significantly transforming the dynamics of flexible work, especially in terms of safety and compliance monitoring. This transformation is evident in various dimensions:

1. **Automated Safety Monitoring**:
   - AI systems enable real-time monitoring of safety standards in gig work, particularly in sectors like transportation and delivery services. For instance, ride-sharing services like Uber utilize AI to monitor driving patterns, ensuring compliance with road safety norms (Henao & Marshall, 2019). This can include detecting unsafe driving behavior and prompting corrective action (Cook, 2020).

2. **Regulatory Compliance**:
   - AI plays a crucial role in ensuring that gig workers and platforms comply with local and international regulations. Automated systems can track and analyze data to ensure adherence to labor laws, tax regulations, and other legal requirements (Prassl, 2018). For instance, platforms like TaskRabbit use AI to help freelancers navigate various tax obligations (De Stefano, 2018).

3. **Worker Health and Safety**:
   - In the context of physical gig jobs, AI-driven tools can monitor worker health and safety conditions. Wearable devices and sensors, integrated with AI, can track vital signs and environmental conditions to prevent accidents and health risks (Leimeister, 2020). For example, construction

gig platforms might use AI to monitor workers' exposure to hazardous conditions (Howard, 2019).

4. **Data Privacy and Protection**:
   - With the increasing use of personal data in gig platforms, AI is crucial for ensuring data privacy and protection. AI algorithms can detect and prevent unauthorized access to personal data, safeguarding both workers' and customers' information (Pasquale, 2015). For example, platforms like Airbnb use AI-driven systems to protect user data and prevent fraud (Guttentag, 2015).

5. **Ethical Standards Monitoring**:
   - AI systems can also monitor and enforce ethical standards within gig platforms. This includes detecting discriminatory practices or biases in job allocations and ensuring fair treatment of all workers (Crawford & Calo, 2016). Platforms like Upwork may employ AI algorithms to detect and prevent bias in job recommendations (Ticona & Mateescu, 2018).

6. **Automated Incident Reporting and Response**:
   - AI enables quick and efficient reporting of safety incidents or compliance violations. Automated reporting systems can prompt immediate responses, thereby mitigating risks and enhancing overall safety standards (Kingsley et al., 2015). For example, delivery apps like DoorDash use AI to report and respond to any incidents during delivery (Möhlmann, 2016).

Challenges and opportunities in the AI-augmented gig economy:

The integration of Artificial Intelligence (AI) into the gig economy presents a complex landscape of challenges and opportunities. This dual impact is significant in various aspects:

1. Opportunity: Enhanced Efficiency and Flexibility:

The opportunity for enhanced efficiency and flexibility is a prominent aspect of the AI-augmented gig economy. This transformation is characterized by several key elements:

1. **Optimized Job Matching**:
   - AI algorithms significantly improve the process of matching gig workers with suitable jobs. Platforms like Upwork and Fiverr utilize AI to analyze vast amounts of data about both freelancers and job postings, facilitating a more precise and efficient pairing (De Stefano, 2018). This optimization not only saves time for both parties but also increases the likelihood of successful job outcomes.
2. **Automated Administrative Tasks**:
   - AI tools help in automating various administrative tasks, such as scheduling, billing, and communication. This automation reduces the time gig workers spend on non-revenue-generating activities, allowing them to focus more on their core competencies and client work (Manyika et al., 2016). For instance, AI-powered chatbots can handle routine client inquiries, while smart scheduling tools can organize appointments and deadlines.
3. **Dynamic Pricing Models**:
   - AI enables more dynamic and market-responsive pricing models. Algorithms can analyze market trends, demand fluctuations, and individual worker performance to suggest optimal pricing strategies. This dynamic approach helps gig workers remain competitive and maximize their earnings (Chen et al., 2015).
4. **Adaptive Learning and Personalization**:
   - AI systems are capable of adaptive learning, meaning they can continuously improve their recommendations and

strategies based on past outcomes. This adaptive approach leads to a more personalized experience for gig workers, as the AI learns from their preferences, skills, and work history to recommend tailored job opportunities and strategies (Agrawal et al., 2018).

5. **Enhanced Project Management**:

   - AI-driven project management tools aid in streamlining workflow and improving time management. These tools can predict project timelines, allocate resources efficiently, and identify potential bottlenecks before they occur, thereby enhancing overall productivity (Bughin et al., 2018).

6. **Real-Time Data Analysis and Decision-Making**:

   - With AI, gig workers can access real-time data and analytics, aiding in informed decision-making. For instance, AI can provide insights into market demand, client feedback, and performance metrics, enabling workers to adjust their strategies proactively (Davenport & Ronanki, 2018).

7. **Facilitating Remote and Flexible Working**:

   - AI technologies are instrumental in supporting remote and flexible working arrangements. Tools like virtual assistants, AI-driven communication platforms, and cloud-based collaboration software make it easier for gig workers to operate from anywhere, thereby expanding their work opportunities (Horton & Zeckhauser, 2016).

II. Challenge: Job Security and Stability:

One of the significant challenges in the AI-augmented gig economy pertains to job security and stability. This issue is multi-faceted and involves several critical aspects:

1. **Fluctuating Work Availability**:

- In the gig economy, work availability can be highly unpredictable, and AI-driven platforms may exacerbate this instability. Algorithms often prioritize matching tasks based on immediate demand, which can result in inconsistent work opportunities for freelancers (Wood et al., 2019). This unpredictability makes it challenging for gig workers to forecast their income and manage their financial stability.

2. **Dependence on Platform Algorithms**:

   - Gig workers are increasingly dependent on platform algorithms for job allocation. These algorithms, which are often opaque, determine the visibility and ranking of workers on the platform, significantly impacting their ability to secure consistent work (Rosenblat & Stark, 2016). This dependence can lead to a lack of control over their career trajectory and uncertainty regarding job continuity.

3. **Reduced Bargaining Power and Worker Protections**:

   - The gig economy, augmented by AI, often classifies workers as independent contractors rather than employees. This classification can lead to a lack of traditional worker protections such as health insurance, pension plans, and unemployment benefits, making their financial and job security more precarious (De Stefano, 2016). Additionally, the dispersed and individualized nature of gig work can reduce collective bargaining power, further affecting job stability.

4. **Competition and Market Saturation**:

   - AI-enhanced platforms can lead to an influx of workers in the gig economy, intensifying competition and potentially driving down wages (Prassl, 2018). Increased competition can make it difficult for individuals to secure enough work to sustain a stable income, particularly for those without specialized skills or established reputations.

5. **Skill Obsolescence and Automation Risks**:
   - The rapid pace of technological change, including advancements in AI, poses a risk of skill obsolescence for gig workers. Skills that are in demand today may become redundant tomorrow due to automation and changing market needs (Acemoglu & Restrepo, 2018). This dynamic requires gig workers to continually update their skills, adding to the instability and uncertainty of their career paths.

6. **Psychological Impact of Job Insecurity**:
   - The lack of job security in the AI-augmented gig economy can have significant psychological impacts, including stress and anxiety related to income unpredictability and future employability (Kalleberg & Vallas, 2018). This stress can affect the overall well-being and work-life balance of gig workers.

7. **Potential for Discrimination and Bias**:
   - AI algorithms, if not carefully designed and monitored, can perpetuate biases in job allocation, potentially discriminating against certain groups of workers based on their data profiles (Sundararajan, 2017). This discrimination can lead to unequal access to job opportunities and further destabilize job security for marginalized groups.

III. Opportunity: Access to Global Markets:

The integration of AI into the gig economy presents significant opportunities, notably in terms of access to global markets. This expansion has several key dimensions:

1. **Breaking Geographical Barriers**:
   - AI-driven platforms in the gig economy allow workers to access job opportunities beyond their local geography, thus breaking traditional barriers to employment (Malhotra & Van Alstyne, 2014). This global reach enables workers

from various regions to compete in a broader marketplace, offering their services to an international client base.

2. **Market Expansion for Freelancers and Small Businesses**:
   - The gig economy, bolstered by AI, provides freelancers and small business owners the opportunity to tap into markets that were previously inaccessible due to geographical and resource limitations (Burtch et al., 2018). This expansion not only increases their potential client base but also allows for diversification of income sources.

3. **Enhanced Visibility and Marketing**:
   - AI algorithms can enhance the visibility of gig workers to a global audience by optimizing their profiles and services based on market trends and search patterns (Horton & Zeckhauser, 2016). This targeted visibility helps freelancers and independent contractors reach a wider, more diverse client base, ultimately leading to more job opportunities.

4. **Cross-Border Collaboration and Innovation**:
   - The global reach of the gig economy fosters cross-border collaborations, leading to a diverse exchange of ideas and innovation (Agrawal et al., 2015). This diversity can spur creativity and allow gig workers to engage in projects that are varied and intellectually stimulating, contributing to professional growth and development.

5. **Language and Cultural Integration**:
   - AI tools, such as language translation and cultural context algorithms, enable workers to better communicate and collaborate with clients from different parts of the world (Brynjolfsson & McAfee, 2014). These tools help overcome language and cultural barriers, making the global marketplace more accessible and inclusive.

6. **Economic Empowerment in Developing Regions**:

- For workers in developing countries, access to global markets through the gig economy can be economically empowering. It offers them the opportunity to earn higher wages compared to local standards and improve their living conditions (Graham et al., 2017).

7. **Learning and Skill Development**:
   - Exposure to global markets and diverse projects enables gig workers to continuously learn and adapt to different work requirements and cultural contexts, enhancing their skills and employability (Kuek et al., 2015).

8. **Data-Driven Market Insights**:
   - AI's ability to analyze vast amounts of data can provide gig workers with valuable insights into global market trends, demand patterns, and client preferences, helping them to strategically position their services (Farrell & Greig, 2016).

IV. Challenge: Skills Gap and Training Needs:

The advent of AI in the gig economy presents not only opportunities but also significant challenges, particularly in addressing the skills gap and training needs. This aspect has multiple dimensions:

1. **Rapid Technological Changes and Skill Obsolescence**:
   - The rapid evolution of AI and related technologies often leads to skill obsolescence, requiring gig workers to continually update their skills (Bughin et al., 2018). This constant need for upskilling can be challenging, especially for those who lack access to training resources or the time to invest in learning.

2. **Access to Training and Educational Resources**:
   - There is a disparity in access to training and educational resources among gig workers. Those in lower-income brackets or in regions with limited educational infrastructure find it more challenging to acquire new skills

necessary for adapting to AI-driven platforms (Katz & Krueger, 2017).

3. **Digital Literacy and Technical Proficiency**:
    - The effective use of AI in the gig economy requires a baseline level of digital literacy and technical proficiency. Workers without these foundational skills may find themselves at a disadvantage in a market increasingly dominated by AI (Autor, 2015).

4. **Cost of Training and Education**:
    - The cost associated with training and education can be prohibitive for many gig workers, particularly independent contractors without the backing of larger organizations. This financial barrier can exacerbate the skills gap (Kaplan & Haenlein, 2019).

5. **Customized Training Programs**:
    - The diverse nature of tasks and roles in the gig economy necessitates customized training programs. However, creating and accessing such tailored training can be challenging, as it requires both resources and an understanding of the specific skill sets needed in various gig roles (Susskind & Susskind, 2015).

6. **Adaptation to AI-Driven Work Environments**:
    - Adapting to AI-driven work environments involves not only technical skills but also soft skills like adaptability, problem-solving, and emotional intelligence. Training programs often overlook these aspects, making it difficult for workers to fully adapt to AI-augmented work settings (Rifkin, 2014).

7. **Inequality in Skill Development Opportunities**:
    - There is a risk of increasing inequality in skill development opportunities. Those with existing higher skill levels and better resources are more likely to benefit from AI,

widening the gap between skilled and less-skilled workers (Acemoglu & Restrepo, 2018).

8. **Collaboration Between Stakeholders**:

- ○ Addressing the skills gap requires collaboration between various stakeholders, including educational institutions, governments, and platform providers. Developing policies and programs that support lifelong learning and skill development is crucial (Manyika et al., 2017).

V. Opportunity: Improved Work-Life Balance:

The integration of AI into the gig economy has opened up opportunities for improved work-life balance, an aspect increasingly valued in modern work environments. This development manifests in several ways:

1. **Flexible Scheduling**:

- ○ AI-driven platforms enable gig workers to have greater control over their work schedules, allowing them to balance work with personal commitments more effectively (De Stefano, 2016). This flexibility is particularly beneficial for those with caregiving responsibilities or pursuing education alongside work.

2. **Remote Work Opportunities**:

- ○ AI has facilitated a rise in remote work opportunities, reducing the need for commuting and allowing individuals to work from locations that suit their lifestyle needs (Katz & Krueger, 2017). This flexibility can significantly enhance the quality of life and reduce stress related to traditional work environments.

3. **Efficient Task Matching**:

- ○ Advanced AI algorithms can match gig workers with tasks that align with their skills and preferred working hours, contributing to a more balanced work life. This efficient matching reduces downtime and increases productivity, allowing for more leisure time (Prassl, 2018).

4. **Reduced Administrative Burdens**:
   - AI applications can automate administrative tasks such as billing, scheduling, and customer communication, freeing up time for gig workers to focus on their core work or personal activities (Manyika et al., 2017).

5. **Personalized Workload Management**:
   - AI systems can analyze a worker's workload and suggest optimal working hours, helping to prevent burnout and ensure a sustainable work pace (Susskind & Susskind, 2015).

6. **Enhanced Career Autonomy**:
   - The gig economy, augmented by AI, empowers individuals to take greater control of their career paths, choosing projects that align with their personal and professional goals (Harris & Krueger, 2015).

7. **Mental Health and Well-being**:
   - Improved work-life balance is linked to better mental health and overall well-being. The autonomy and flexibility provided by AI-driven gig work can lead to higher job satisfaction and reduced stress levels (De Stefano, 2016).

8. **Challenges in Consistency and Income Stability**:
   - While AI in the gig economy can improve work-life balance, it also presents challenges in terms of consistent work and income stability, which are crucial components of overall life satisfaction (Kaplan & Haenlein, 2019).

VI. Challenge: Data Privacy and Worker Surveillance:

The intersection of AI and the gig economy raises significant challenges concerning data privacy and worker surveillance. These challenges have broad implications:

1. **Increased Data Collection**:
   - AI-driven platforms in the gig economy often necessitate the collection of vast amounts of personal data from workers, including location, work habits, and communication patterns (Möhlmann & Zalmanson, 2017). This extensive data collection raises concerns about the privacy and security of gig workers' personal information.

2. **Surveillance and Monitoring**:
   - AI technology enables continuous monitoring and surveillance of gig workers. While this can optimize efficiency and safety, it also creates a sense of constant oversight, which can be perceived as invasive and stressful for workers (Rosenblat & Stark, 2016).

3. **Algorithmic Management**:
   - Decisions regarding task allocation, performance evaluations, and even terminations are increasingly made by algorithms, often without transparency or recourse for workers (Lee, Kusbit, Metsky, & Dabbish, 2015). This lack of transparency can exacerbate workers' feelings of vulnerability and lack of control over their work environment.

4. **Data Security Risks**:
   - The gig economy platforms, relying heavily on AI, are targets for cyber threats. Breaches in these systems can lead to the exposure of sensitive worker data, posing significant risks to their privacy and security (Kshetri, 2017).

5. **Ethical Considerations**:
   - The ethical implications of using AI for worker surveillance in the gig economy are a growing concern. The balance between efficient management and respect for worker privacy is a critical area of debate (Martin, 2019).

6. **Legal and Regulatory Challenges**:
   - Existing laws and regulations may not adequately address the unique challenges posed by AI-driven surveillance in

the gig economy. This gap necessitates the development of new legal frameworks to protect worker privacy (De Stefano, 2016).

7. **Worker Trust and Morale**:
   - Invasive monitoring and lack of privacy can erode trust between gig workers and platforms, negatively impacting worker morale and potentially leading to decreased productivity (Möhlmann & Zalmanson, 2017).

8. **Potential for Discriminatory Practices**:
   - AI algorithms, if not carefully designed and monitored, can lead to discriminatory practices in surveillance and data handling, raising concerns about fairness and equality in the gig economy (Graham, Hjorth, & Lehdonvirta, 2017).

VII. Opportunity: Creation of New Job Categories:

The integration of AI into the gig economy not only transforms existing job roles but also creates entirely new categories of employment:

1. **Emergence of AI-Driven Roles**:
   - The demand for AI specialists, data analysts, and machine learning engineers in the gig economy is escalating. These roles are crucial for developing and maintaining the AI systems that power gig platforms (Brynjolfsson & McAfee, 2014).

2. **Hybrid Jobs**:
   - There's a growing trend of hybrid jobs that combine traditional skills with tech-savvy capabilities. For instance, digital marketing professionals in the gig economy now need to understand AI-based analytics tools to remain competitive (Manyika et al., 2017).

3. **Micro-Tasks and AI Training**:
   - The AI sector has created a demand for gig workers to perform micro-tasks, such as data labeling and training AI

algorithms. This has led to the emergence of a niche job sector within the gig economy (Gray & Suri, 2019).

4. **Freelance AI Consultants**:
   - There's a growing market for freelance AI consultants who advise businesses on AI integration, demonstrating a fusion of gig work and high-tech consultancy (Kaplan & Haenlein, 2019).

5. **Custom AI Solution Development**:
   - Small businesses and startups increasingly rely on gig workers for the development of bespoke AI solutions, tailored to their specific needs, fostering a new category of entrepreneurial tech roles (Bughin, Hazan, Ramaswamy, Chui, Allas, Dahlström, Henke, & Trench, 2018).

6. **Ethical AI Auditors**:
   - With the rise of AI, there's a burgeoning need for professionals who can audit AI systems for ethical compliance and bias, a role increasingly found in the gig economy (Martin, 2019).

7. **AI-Enhanced Creative Roles**:
   - AI is not just for technical roles; it's also creating new opportunities in creative fields. For example, graphic designers and content creators are using AI tools to enhance their work and offer innovative services (Deng, Joshi, & Zhang, 2018).

8. **Remote AI Education and Training**:
   - The gig economy is seeing a rise in demand for professionals who can provide remote AI education and training, catering to the ongoing need for upskilling in this area (Bughin et al., 2018).

VIII. Challenge: Regulatory and Ethical Concerns:

The integration of AI into the gig economy presents unique regulatory and ethical concerns:

1. **Lack of Standard Regulatory Frameworks**:
   - The gig economy, especially when augmented with AI, operates in a space that often lacks clear regulatory frameworks. This absence of standardization can lead to issues like worker misclassification and unfair labor practices (De Stefano, 2016).

2. **Ethical Use of AI**:
   - The ethical use of AI in gig work raises concerns, particularly regarding privacy, data security, and the potential for bias in AI algorithms. The need for ethical guidelines and oversight is critical to prevent discrimination and protect worker rights (Martin, 2019).

3. **Worker Rights and Protections**:
   - Traditional labor laws are often ill-equipped to address the unique challenges faced by gig workers in an AI-driven marketplace. This includes issues like job security, benefits, and fair compensation (Harris & Krueger, 2015).

4. **Data Privacy Issues**:
   - AI systems rely heavily on data, raising concerns about the privacy and security of both workers and clients in the gig economy. Ensuring data protection while leveraging AI technology is a significant challenge (Pasquale, 2015).

5. **Algorithmic Transparency**:
   - The algorithms used in AI can sometimes be a "black box," lacking transparency. This raises concerns about how decisions are made, especially those affecting worker allocation, performance evaluations, and remuneration (Burrell, 2016).

6. **Responsibility and Accountability**:
   - Determining responsibility for AI-driven decisions in the gig economy is complex. This includes issues of liability in case of errors or accidents involving AI systems (Calo, 2015).

7. **Global Regulatory Variations**:

   - The global nature of the gig economy means that AI integration has to navigate a maze of different legal and ethical standards across countries, complicating compliance and operational strategies (Schmidt, 2017).

8. **Potential for Surveillance and Control**:

   - The use of AI for worker monitoring and management in the gig economy can lead to excessive surveillance, raising ethical concerns about worker autonomy and privacy (Möhlmann & Zalmanson, 2017).

11

Chapter 6: Education and Training in the AI Age

The arrival of Artificial Intelligence (AI) has brought significant changes to education and training, necessitating a reevaluation of traditional methods and curricula. This evolution is essential to prepare individuals for the AI-driven workforce and society.

1. Redefining Skill Sets:

In the AI age, there's a growing emphasis on developing skills like problem-solving, critical thinking, and creativity, alongside technical prowess in AI and machine learning (Brynjolfsson & McAfee, 2014). Traditional educational frameworks are adapting to include these skills, which are essential for working effectively with AI technologies.

The integration of Artificial Intelligence (AI) into various sectors has necessitated a redefinition of skill sets required in the modern workforce. This shift focuses not only on technical skills related to AI but also on a range of soft skills and cognitive abilities.

1. Emphasis on Cognitive and Emotional Skills:

- In an AI-dominated landscape, cognitive skills such as critical thinking, problem-solving, and decision-making become increasingly valuable (Schwab, 2016). Emotional intelligence, including empathy and interpersonal skills, is also crucial in environments where AI handles more routine tasks (Goleman, 1995).

2. **Technical Skills for AI and Machine Learning**:

- Proficiency in AI, machine learning, data analysis, and programming is becoming essential, not just for IT professionals but across various sectors. This includes understanding AI algorithms, data science, and their practical applications (Russell & Norvig, 2016).

3. **Adaptability and Continuous Learning**:

- The rapid evolution of AI technologies demands adaptability and a commitment to continuous learning. This enables workers to stay relevant and adept at using emerging tools and technologies (Brynjolfsson & McAfee, 2014).

4. **Creativity and Innovation**:

- As AI takes over more routine and analytical tasks, human creativity and innovation become key differentiators in the workforce. Education systems are increasingly focusing on fostering creativity and innovative thinking (Florida, 2012).

5. **Interdisciplinary Knowledge**:

- The interconnectedness of AI with various fields necessitates an interdisciplinary approach to education. Understanding the impact of AI on economics, ethics, law, and social sciences is crucial (Bostrom, 2014).

6. **Digital Literacy**:

- Digital literacy, including the ability to interact with and understand digital platforms, is increasingly important. This encompasses not only the use of tools but also an understanding of digital security and ethics (Eubanks, 2018).

7. **Collaborative Skills**:
   - Collaboration skills are vital as many AI projects involve multidisciplinary teams. This includes the ability to work effectively in diverse teams, communicate ideas, and manage projects (Sawyer, 2007).
8. **Ethical and Societal Understanding**:
   - An understanding of the ethical implications of AI and its impact on society is crucial. This includes issues related to bias, privacy, and the broader societal changes brought about by AI (Mittelstadt et al., 2016).

II. AI in Personalized Learning:

AI has the potential to personalize education by adapting learning materials to the individual needs and pace of each student. Systems can provide tailored recommendations and identify areas needing improvement, thereby enhancing learning efficiency (Zawacki-Richter, O., et al., 2019).

The integration of Artificial Intelligence (AI) in education is revolutionizing the concept of personalized learning, tailoring educational experiences to individual learners' needs, abilities, and preferences.

1. **Adaptive Learning Systems**:
   - AI-driven adaptive learning systems analyze student performance in real-time, adjusting the difficulty and style of content to match their learning pace and preferences (Pane et al., 2017). This approach ensures that each student receives a customized learning experience.
2. **Predictive Analytics for Student Success**:
   - AI algorithms can predict student performance and identify those at risk of falling behind. This enables early intervention strategies to be implemented, enhancing student success rates (Xu et al., 2019).
3. **Enhanced Engagement through Gamification**:

- AI facilitates the incorporation of gamification into learning, increasing student engagement and motivation. AI can adapt game-based learning activities to individual learning styles, making education both fun and effective (Hamari et al., 2014).

4. **Facilitating Diverse Learning Styles**:
   - AI technologies support diverse learning styles by providing varied content formats like videos, interactive simulations, and text, catering to visual, auditory, and kinesthetic learners (Fleming & Mills, 1992).

5. **Continuous Feedback and Assessment**:
   - AI systems provide continuous feedback and assessment, enabling learners to understand their progress and areas needing improvement. This immediate feedback loop is crucial for effective learning (Shute & Zapata-Rivera, 2012).

6. **Language Learning and Literacy Enhancement**:
   - AI-driven language learning tools and literacy applications offer personalized guidance, adapting to individual language proficiency levels and learning speeds (Warschauer & Healey, 1998).

7. **Support for Special Education**:
   - AI tools offer significant potential in special education, providing customized learning experiences and support for students with disabilities, catering to their unique educational needs (Lancioni et al., 2012).

8. **Data-Driven Curriculum Development**:
   - AI's ability to analyze vast amounts of educational data can guide curriculum development, ensuring it meets the evolving needs of students and aligns with current industry trends (Baker & Inventado, 2014).

III. Training for AI Literacy:

As AI becomes increasingly ubiquitous, there is a need for widespread AI literacy. This involves understanding AI's capabilities, limitations, and ethical implications, not just for specialists but for the general populace (Russell, S., & Norvig, P., 2016).

Training for AI literacy has become a crucial aspect of education and workforce development in the AI age. This involves equipping individuals with the knowledge and skills necessary to understand, use, and interact with AI technologies effectively.

1. **Curriculum Integration**:
    - Schools and universities are integrating AI concepts into their curricula, providing foundational knowledge in AI and machine learning. This includes understanding algorithms, data analysis, and ethical implications of AI (Hwang, 2020). Such integration prepares students for a workforce increasingly reliant on AI technologies.

2. **Professional Development and Continuous Learning**:
    - For professionals, training programs and workshops focusing on AI literacy are essential for staying current with technological advancements. Continuous learning opportunities in AI are crucial for employees to adapt to rapidly evolving job requirements (Brynjolfsson & Mitchell, 2017).

3. **Ethical and Responsible AI Use**:
    - Training programs also emphasize the ethical use of AI, including understanding biases in AI systems, ensuring privacy, and promoting responsible AI deployment (Mittelstadt et al., 2016). This aspect of AI literacy is vital for both developers and users.

4. **Practical AI Applications**:
    - Hands-on experience with AI tools and platforms in various sectors such as healthcare, finance, and manufacturing

allows learners to understand the practical applications and limitations of AI (Davenport & Ronanki, 2018).

5. **Collaborative Skills for Human-AI Interaction**:

   ○ As AI becomes more prevalent in workplaces, training programs focus on collaborative skills necessary for effective human-AI interaction, including teamwork, communication, and adaptability (Daugherty & Wilson, 2018).

6. **Critical Thinking and Problem-Solving in AI Contexts**:

   ○ Training programs are increasingly emphasizing critical thinking and problem-solving skills within AI contexts, enabling learners to make informed decisions when working with AI systems (Bostrom & Yudkowsky, 2014).

7. **AI Literacy for All Ages**:

   ○ Recognizing the importance of AI literacy across all age groups, educational initiatives are targeting not only young students but also older adults, ensuring inclusive access to AI education (Tegmark, 2017).

8. **Interdisciplinary Approach**:

   ○ AI literacy training often adopts an interdisciplinary approach, combining insights from computer science, sociology, psychology, and philosophy, to provide a holistic understanding of AI's impact on society (Russell & Norvig, 2016).

IV. Ethical and Societal Implications:

Education systems are also integrating discussions about the ethical and societal implications of AI. This includes addressing issues like privacy, bias, and job displacement, preparing individuals to make informed decisions about AI usage (Eubanks, V., 2018).

Education and training in the age of AI must address not only the technical skills required to develop and use AI systems but also the ethical and societal implications of these technologies. This holistic approach is crucial in preparing individuals to navigate the complexities associated with AI.

1. **Understanding AI Ethics**:
    - Courses and training modules increasingly focus on AI ethics, including topics like algorithmic bias, data privacy, and the ethical design and deployment of AI systems (Mittelstadt et al., 2016). These educational efforts are critical for ensuring that AI is used in a manner that respects human rights and values.

2. **Impact on Society and Workforce**:
    - Education programs discuss the impact of AI on society and the workforce, examining issues like job displacement, the changing nature of work, and the skills gap. These discussions prepare individuals for the evolving job market and promote awareness of how AI can both create and eliminate job opportunities (Brynjolfsson & McAfee, 2014).

3. **Promoting Responsible Innovation**:
    - Training in responsible innovation is essential, emphasizing the importance of considering the societal impacts of AI technologies during their development and deployment (Stahl, 2013). This includes assessing potential risks and benefits and engaging with diverse stakeholders.

4. **Cultural and Social Sensitivity**:
    - AI education includes cultural and social sensitivity training to ensure that AI systems are inclusive and do not perpetuate existing societal biases or inequalities (Adam, 2005). This is especially important given the global reach of AI technologies.

5. **Global Perspectives on AI Governance**:
    - Understanding different global perspectives on AI governance, including various regulatory and policy approaches,

is a key component of AI education. This helps in comprehending how AI is managed in different cultural and legal contexts (Cath et al., 2018).

6. **Public Awareness and Engagement**:
   - Education and training programs also aim to enhance public awareness and engagement with AI. This involves demystifying AI and encouraging informed public discourse about its benefits and challenges (Bostrom & Yudkowsky, 2014).

7. **Interdisciplinary Learning**:
   - AI education adopts an interdisciplinary approach, incorporating insights from fields such as philosophy, sociology, and law, to provide a comprehensive understanding of the multifaceted impact of AI on society (Russell & Norvig, 2016).

V. Lifelong Learning and Reskilling:

The rapid evolution of AI technologies necessitates a focus on lifelong learning and reskilling, especially for those in industries most impacted by automation. Educational institutions and corporate training programs are increasingly offering courses for reskilling and upskilling in AI-related areas (Schwab, K., 2017).

In the age of AI, the concept of lifelong learning and reskilling becomes increasingly important. As AI continues to advance, it is reshaping job markets and the skills required, necessitating continuous education and skill development.

1. **Adapting to Technological Changes**:
   - Lifelong learning is essential for adapting to the rapid technological changes brought about by AI. Workers must continuously update their skills to remain relevant in an evolving job market (Schwab, 2017). This includes both

technical skills related to AI and complementary skills such as critical thinking and creativity.

2. **Reskilling Initiatives**:
   - Many organizations and governments have launched re-skilling initiatives to address the skills gap created by AI advancements. These programs focus on training workers in areas most likely to be affected by automation and AI (Bughin et al., 2018). The aim is to transition workers into new roles that AI is less likely to replace.

3. **Online Learning Platforms**:
   - The rise of online learning platforms has made education more accessible, allowing individuals to learn new skills at their own pace and convenience. These platforms offer courses in AI, data science, and other relevant fields, making it easier for workers to acquire the skills needed for the AI age (Huang et al., 2020).

4. **Corporate Training Programs**:
   - Companies are increasingly investing in training programs to help their employees adapt to new technologies. These programs often focus on digital literacy, AI fundamentals, and ways AI can be applied within their specific roles (Capgemini, 2019).

5. **Government Policies and Support**:
   - Governments play a crucial role in promoting lifelong learning and reskilling. Policies and funding support are essential for developing national education programs that can cater to the needs of a workforce impacted by AI (OECD, 2019).

6. **Collaboration between Industry and Academia**:
   - Partnerships between industry and academia are crucial for aligning education and training programs with the actual demands of the job market. Such collaborations can

ensure that training programs are relevant and up-to-date (World Economic Forum, 2018).

7. **Cultivating a Culture of Continuous Learning**:
   - Creating a culture that values and encourages continuous learning is vital. This involves changing mindsets to view learning as an ongoing process rather than a one-time event (Mehta, 2017).

8. **Focus on Soft Skills**:
   - Alongside technical skills, there is a growing emphasis on soft skills like problem-solving, emotional intelligence, and adaptability. These skills are critical in a landscape where AI tools handle more routine tasks (Deming, 2017).

VI. Collaboration between Academia and Industry:

Partnerships between educational institutions and the tech industry are vital to ensure that curricula remain relevant and up-to-date with the latest AI advancements. These collaborations can also provide practical experience through internships and projects (Manyika, J., et al., 2017).

The intersection of academia and industry plays a pivotal role in shaping education and training in the AI age. This collaboration is essential for aligning educational curricula with industry needs, ensuring that learners are equipped with relevant skills for the evolving job market.

1. **Developing Relevant Curricula**:
   - Collaboration between academia and industry can lead to the development of curricula that are directly aligned with the current and future needs of the job market. Industry partners can provide insights into the skills and knowledge required in the AI-driven workforce, ensuring that academic programs are relevant and up-to-date (Brynjolfsson & McAfee, 2014).

2. **Real-World Experience:**
   - Partnerships can facilitate internships, co-op programs, and apprenticeships that provide students with valuable real-world experience. These opportunities allow students to apply their theoretical knowledge in practical settings, bridging the gap between education and employment (Hagel et al., 2016).

3. **Joint Research and Development:**
   - Joint research initiatives between academia and industry can lead to innovative developments in AI. These collaborations can result in cutting-edge research, driving both academic inquiry and industrial advancement (Agrawal et al., 2018).

4. **Lifelong Learning and Professional Development:**
   - Industry collaboration can support lifelong learning and professional development programs for existing employees. By understanding the skills gap in their workforce, companies can work with academic institutions to create tailored training programs (WEF, 2018).

5. **Adaptability to Technological Changes:**
   - The dynamic nature of AI technology requires a workforce that can quickly adapt to changes. Collaborative programs can help in continuously updating the skills of both new entrants and existing workers in the job market (Autor, 2015).

6. **Enhanced Teaching Methodologies:**
   - Input from industry can help academia in adopting more practical and applied teaching methodologies. This includes project-based learning, simulation-based training, and the use of real-world case studies (Rothwell et al., 2019).

7. **Funding and Resource Sharing:**

- ○ Collaboration can also involve resource sharing, including funding for research and development, as well as access to advanced tools and technologies. This can enhance the quality of education and research in AI (Lee et al., 2020).

8. **Policy Development and Advocacy**:

- ○ Joint efforts can influence policy development related to AI education and training. By working together, academia and industry can advocate for policies that support the growth and development of AI talent (Susskind & Susskind, 2015).

VII. Online and Distance Learning:

AI is facilitating the growth of online and distance learning platforms, making education more accessible to a broader audience. AI-driven analytics can enhance the effectiveness of these platforms by monitoring student engagement and performance (Ferguson, R., & Clow, D., 2017).

Online and distance learning have become integral components of education in the AI age, reshaping how knowledge is delivered and accessed globally. This shift has significant implications for educational accessibility, pedagogy, and learner engagement.

1. **Increased Accessibility and Flexibility**:

- ○ Online learning platforms have democratized access to education, allowing learners from diverse backgrounds and geographical locations to access quality educational resources (Bates, 2015). These platforms offer flexibility in learning, accommodating different learning styles and schedules (Allen & Seaman, 2014).

2. **Use of AI for Personalized Learning**:

- ○ AI technologies in online education can personalize learning experiences, adapting content and pacing to individual learner needs (Zhou et al., 2020). This approach enhances learner engagement and outcomes by providing tailored educational paths.

3. **Enhanced Interaction and Collaboration Tools**:

    ○ Online learning environments incorporate tools for real-time interaction and collaboration, facilitating peer-to-peer learning and engagement. Features like forums, video conferencing, and collaborative projects enable a community-based learning experience (Means et al., 2013).

4. **Scalability and Resource Efficiency**:

    ○ Distance learning models offer scalability, allowing institutions to reach a larger number of students without the constraints of physical space and resources. This scalability is essential in disseminating education in a cost-effective manner (Daniel, 2012).

5. **Integration of Multimedia and Interactive Content**:

    ○ Digital learning environments leverage multimedia content, including videos, animations, and interactive simulations, to enrich the learning experience. Such content caters to different learning styles and enhances comprehension (Mayer, 2014).

6. **Assessment and Feedback through AI**:

    ○ AI-driven assessment tools in online learning can provide immediate, personalized feedback to learners, a critical aspect of the learning process. These tools can assess learner performance and provide targeted feedback to improve learning outcomes (Buckingham Shum & Ferguson, 2012).

7. **Challenges in Engagement and Motivation**:

    ○ Despite its benefits, online learning faces challenges in student engagement and motivation. The lack of physical presence can lead to feelings of isolation and decreased motivation among learners (Hrastinski, 2008).

8. **Digital Divide and Accessibility Concerns**:

- The reliance on technology in online learning highlights the digital divide, where students without reliable internet access or technology are at a disadvantage (Selwyn, 2014).

9. **Continuous Evolution of Educational Technology**:

- The field of online learning is continually evolving, with new technologies and approaches emerging regularly. This requires educators and institutions to stay updated and adapt to these changes (Veletsianos & Shepherdson, 2016).

VIII. Preparing for Future Job Markets:

Education systems are increasingly focusing on preparing students for future job markets that will be heavily influenced by AI and automation. This involves not just technical training but also fostering adaptability and resilience (World Economic Forum, 2018).

The rapid advancements in AI and technology have significant implications for the future job market, necessitating a reevaluation and adaptation of educational and training programs. This preparation involves developing new skill sets, fostering adaptability, and anticipating future industry needs.

1. **Emphasis on Digital Literacy and Technical Skills**:

- In the AI age, digital literacy becomes a foundational skill. Educational institutions must integrate technology and AI-related subjects into their curricula to prepare students for a tech-centric job market (van Laar et al., 2017). This includes programming, data analysis, and understanding AI and machine learning fundamentals.

2. **Soft Skills and Critical Thinking**:

- While technical skills are crucial, the importance of soft skills like critical thinking, creativity, and problem-solving is amplified in the AI age. These skills enable individuals to adapt to new challenges and technological changes that AI brings to the workplace (Bughin et al., 2018).

3. **Interdisciplinary Education**:
   - The future job market demands an interdisciplinary approach to education. Combining AI and technology education with fields like humanities, social sciences, and business prepares students for diverse roles and enhances their adaptability (Fiala et al., 2019).

4. **Continuous Learning and Upskilling**:
   - As AI continues to evolve, the need for continuous learning and upskilling becomes essential. Lifelong learning programs and online courses offer opportunities for professionals to stay current with technological advancements (Schwab & Samans, 2016).

5. **Ethical and Societal Implications**:
   - Understanding the ethical and societal implications of AI is crucial. Education programs must incorporate discussions about AI ethics, data privacy, and the societal impact of automation and AI technologies (Coeckelbergh, 2020).

6. **Career Guidance and AI Awareness**:
   - Career guidance services should include information about emerging AI-driven industries and the evolving nature of various professions. This helps students and professionals make informed decisions about their career paths in the context of an AI-influenced job market (Brynjolfsson & McAfee, 2014).

7. **Collaboration with Industry**:
   - Educational institutions should collaborate with industry partners to align curricula with the skills required in the job market. Such partnerships can also provide practical experience through internships and project-based learning (Hernández-de-Menéndez et al., 2019).

8. **Adapting to Job Displacement**:
   - As AI and automation may lead to job displacement in certain sectors, educational programs need to prepare

individuals for transitions into new roles or industries. This includes retraining and support for displaced workers (Acemoglu & Restrepo, 2020).

9. **Emphasizing Entrepreneurship**:

   ○ Encouraging entrepreneurship and innovation in education can prepare individuals to create and capitalize on new opportunities in the AI-driven economy (Audretsch & Link, 2019).

Evolving educational curricula for the future:

As AI continues to reshape the landscape of work and society, the evolution of educational curricula becomes imperative to prepare future generations for the challenges and opportunities that lie ahead. This evolution involves integrating AI and technology into learning processes, fostering new skill sets, and aligning education with future job market demands.

1. **Integration of AI and Emerging Technologies in Curricula**:

Incorporating AI and emerging technologies into educational curricula is essential. This integration should not be limited to STEM fields but extended across various disciplines to provide a holistic understanding of AI's impact (Luckin et al., 2016). Courses on AI, machine learning, and data science should become standard offerings in higher education.

The integration of artificial intelligence (AI) and emerging technologies into educational curricula is a critical step towards preparing students for a future increasingly shaped by these advancements. This integration involves not just adding new courses, but rethinking the way subjects are taught and the skills that are emphasized.

1. **Curriculum Design for AI and Technology**:

- Educational institutions must redesign curricula to include AI and technology-focused subjects. This involves not just the addition of new courses but integrating these concepts into existing subjects to create a more holistic understanding (Fadel, 2016). For example, incorporating AI in health education could involve teaching how machine learning can be used in medical diagnostics.

2. **Interdisciplinary Approach**:

   - Integrating AI into the curriculum should be approached in an interdisciplinary manner. This means that AI is not only taught in computer science departments but is also integrated into courses like business, healthcare, and social sciences, demonstrating its wide-ranging applications (Luckin et al., 2016).

3. **Hands-on Learning and Practical Applications**:

   - Practical, hands-on learning is crucial for students to understand and apply AI concepts. This could involve lab work, projects, or collaborations with tech companies, providing real-world experience and understanding of how AI is used in various industries (Hernández-de-Menéndez et al., 2019).

4. **Updating Faculty Skills and Knowledge**:

   - To effectively teach these new curricula, educators themselves need to be up-to-date with AI and technology. Professional development and ongoing training for teachers are essential, ensuring they have the necessary skills and knowledge to guide students effectively (Conley, 2019).

5. **Incorporating AI Tools in Teaching**:

   - Using AI tools in the teaching process can provide a practical demonstration of the technology's capabilities. This includes using AI for personalized learning, automated grading, and enhancing classroom engagement (Zawacki-Richter et al., 2019).

6. **Ethical and Societal Implications**:
   - Any curriculum on AI must include a component on ethics. Students should learn about the ethical use of AI, bias in machine learning, privacy concerns, and the broader societal impacts of technology (Coeckelbergh, 2020).
7. **Global Perspective**:
   - The curriculum should provide a global perspective on AI, highlighting different approaches, applications, and implications in various cultural and geographical contexts. This helps students appreciate the global impact of AI technologies (Reimers & Chung, 2016).
8. **Flexibility and Adaptability**:
   - Given the rapid advancement in AI and technology, curricula need to be flexible and adaptable, allowing for the integration of the latest developments and discoveries in the field (Schwab & Samans, 2016).

II. Focus on Digital Competence and Computational Thinking:

Developing digital competence is critical in the AI age. This includes not only basic IT skills but also computational thinking, which involves problem-solving skills that leverage the capabilities of AI and computer systems (Voogt et al., 2015).

In the age of AI, evolving educational curricula to include a strong focus on digital competence and computational thinking is essential. These skills are not just for those pursuing careers in technology but are increasingly important across all sectors.

1. **Digital Competence as a Core Skill**:
   - In the digital age, it's critical that educational curricula emphasize digital competence, which includes not only the ability to use digital tools but also an understanding of their impact on society and individuals (Van Laar et al.,

2017). This involves teaching students how to navigate the digital world, critically assess information, and understand online safety and digital ethics.

2. **Integrating Computational Thinking**:

    ◦ Computational thinking, a problem-solving process involving skills like pattern recognition, abstraction, and algorithm design, is vital in understanding and applying AI technologies. It's important that curricula integrate computational thinking across different subjects, not just computer science (Wing, 2006). This approach can enhance students' problem-solving skills in various disciplines.

3. **Project-Based Learning**:

    ◦ Implementing project-based learning that incorporates digital tools and computational thinking can provide students with practical, hands-on experience. Such projects encourage creativity, collaboration, and critical thinking, essential skills in the AI-driven world (Bell, 2010).

4. **Cross-Disciplinary Approach**:

    ◦ Digital competence and computational thinking should be incorporated into a wide range of subjects. For example, a history class might involve using digital archives and databases, while a science class might use computational models and simulations (Voogt et al., 2015).

5. **Teacher Training and Professional Development**:

    ◦ To effectively teach these skills, educators need professional development and resources. Continuous training is essential for teachers to stay abreast of technological advancements and integrate these skills into their teaching (Honey & Kanter, 2013).

6. **Focus on Critical Thinking and Ethics**:

    ◦ As students learn digital skills, it's equally important to emphasize critical thinking and ethical considerations related to technology use, including discussions about data privacy, misinformation, and the societal impact of technology (O'Neil, 2016).

7. **Preparing for Digital Citizenship**:
    - Curricula should aim to prepare students not just for the workforce but also for their role as digital citizens. This includes understanding the responsibilities and ethical considerations that come with participating in the digital world (Ribble, 2015).

8. **Adapting to Evolving Technologies**:
    - Educational curricula must be adaptable to keep pace with rapidly evolving technologies. This flexibility ensures that students are always learning skills relevant to the current and future technology landscape (Bower, 2017).

III. Emphasis on Soft Skills and Adaptability:

Soft skills such as critical thinking, creativity, emotional intelligence, and adaptability are increasingly important. These skills enable individuals to work alongside AI and adapt to rapidly changing work environments (Bughin et al., 2018).

The evolving educational curricula in the age of AI must place a significant emphasis on soft skills and adaptability. In a rapidly changing technological landscape, these skills are increasingly recognized as vital for success in both personal and professional domains.

1. **Soft Skills as Fundamental Competencies**:
    - Soft skills, such as communication, teamwork, emotional intelligence, and problem-solving, are becoming increasingly important in the workforce. In the context of AI, where technical skills can often be automated, soft skills differentiate human workers and add value that AI cannot replicate (Deming, 2017). These skills enable individuals to navigate complex social environments and collaborate effectively.

2. **Integrating Soft Skills into Curricula**:
   - It's important for educational institutions to integrate soft skills training into their curricula actively. This can be achieved through group projects, presentations, and role-playing exercises, which help students develop communication, leadership, and teamwork abilities (Binkley et al., 2012).

3. **Adaptability and Continuous Learning**:
   - In a world where technological advancements rapidly make specific skills obsolete, the ability to adapt and engage in lifelong learning is crucial. Educational systems should foster adaptability by teaching students how to learn, unlearn, and relearn as necessary (Mehta & Aguilera, 2018).

4. **Critical Thinking and Problem Solving**:
   - Critical thinking and problem-solving skills are essential in navigating the challenges posed by AI and automation. These skills enable individuals to analyze information critically, make informed decisions, and solve complex problems (Trilling & Fadel, 2009).

5. **Cultural Competence and Global Awareness**:
   - As the world becomes more interconnected, cultural competence and global awareness become essential soft skills. Educational curricula should include opportunities to understand and appreciate diverse cultures and perspectives, preparing students for a globalized workplace (Hunter et al., 2006).

6. **Ethical Decision-Making**:
   - In the AI age, ethical considerations are paramount. Education should include a focus on ethical decision-making, teaching students to consider the broader impacts of technology on society and the environment (Bryson, 2018).

7. **Embracing Change and Resilience**:

- Curricula should help students develop resilience and an openness to change. This can be fostered through exposure to new ideas, encouraging creativity, and teaching strategies for coping with failure and setbacks (Zimmerman, 2013).

8. **Personalized Learning Paths**:
- Recognizing that soft skills development varies among individuals, personalized learning paths can be effective. Leveraging AI in education to tailor learning experiences can help students develop these skills at their own pace (Pane et al., 2017).

IV. Interdisciplinary and Project-Based Learning:

Education should encourage interdisciplinary learning, blending technology with arts, humanities, and social sciences. Project-based learning can also help students apply AI and technological concepts in real-world scenarios (Conley, 2019).

The evolution of educational curricula in the AI age necessitates a shift towards interdisciplinary and project-based learning (PBL). This approach aligns with the need for diverse skills and knowledge in a technology-driven world.

1. **Interdisciplinary Learning for a Holistic Education**:
- Interdisciplinary learning, which integrates concepts and perspectives from different disciplines, is crucial in an AI-dominated landscape. It encourages a holistic understanding of problems, combining technical knowledge with insights from social sciences, humanities, and arts (Jacobs, 2013). This approach reflects the interconnected nature of modern challenges and solutions, especially in AI-related fields.

2. **Project-Based Learning for Real-World Application**:

- PBL is a dynamic classroom approach where students actively explore real-world problems and challenges. This method has been shown to enhance problem-solving and critical thinking skills, which are essential in an AI-driven workforce (Thomas, 2000). PBL also encourages innovation and creativity, vital for navigating the complexities of AI technologies.

3. **Enhancing Collaboration and Teamwork**:

   - Both interdisciplinary and PBL approaches naturally foster collaboration and teamwork. In an AI-centric future, the ability to work effectively in diverse teams becomes more important. These educational strategies teach students how to integrate different viewpoints and collaborate on complex projects (Laal & Laal, 2012).

4. **Incorporating AI and Technology in Projects**:

   - Integrating AI and technology into project-based learning allows students to gain hands-on experience with these tools. This experience is invaluable in preparing them for future careers where AI will be ubiquitous (Hwang, 2014).

5. **Developing Adaptive and Flexible Thinking**:

   - By engaging with a variety of disciplines and tackling real-world problems, students develop adaptive and flexible thinking. This is critical in an era where change is the only constant, and the ability to adapt is key to success (Wagner, 2008).

6. **Building Research and Inquiry Skills**:

   - Interdisciplinary and PBL approaches enhance research and inquiry skills. These skills are fundamental in an information-rich world, allowing learners to navigate and analyze vast amounts of data, a common scenario in AI-related tasks (Krajcik & Blumenfeld, 2006).

7. **Emphasizing Continuous Learning and Curiosity**:

- These approaches instill a love for learning and intellectual curiosity, essential for lifelong learning. With AI continually evolving, the ability to continuously learn and stay curious is invaluable (Dweck, 2006).

8. **Addressing Ethical and Societal Impacts**:

- Interdisciplinary learning particularly allows for the exploration of ethical and societal impacts of AI, a critical aspect of responsible AI use and development (Eckhardt et al., 2018).

V. Ethics, Privacy, and Societal Impact of AI:

Curricula must include discussions on the ethical implications, privacy concerns, and the broader societal impact of AI. This education is crucial for developing responsible AI use and understanding its consequences (Coeckelbergh, 2020).

The integration of AI in education necessitates a focus on the ethical, privacy, and societal implications of this technology. This is critical to ensure responsible use and understanding of AI systems.

1. **Ethics in AI**:

- Ethical considerations in AI are paramount. Educational curricula should include topics like algorithmic bias, fairness, and the ethical use of AI. Students need to be aware of how AI decisions are made and the moral implications of these decisions (Mittelstadt et al., 2016). Ethics in AI education prepares students to develop and use AI responsibly, ensuring technologies are designed and implemented with societal values in mind.

2. **Privacy Concerns**:

- With AI's ability to process vast amounts of personal data, privacy becomes a significant concern. Education systems must address the importance of data privacy and security, teaching students about data protection laws, consent, and data governance (Polonetsky & Tene, 2014). Understanding these concepts is crucial in an era where personal data is a valuable commodity.

3. **Societal Impact of AI**:

    ○ The societal impact of AI, including its effects on employment, social interactions, and democracy, should be a key component of modern curricula. Discussions around how AI can both positively and negatively impact society help students understand the broader implications of this technology (Brynjolfsson & McAfee, 2014). This understanding is essential for developing AI solutions that benefit society as a whole.

4. **Case Studies and Real-World Examples**:

    ○ Incorporating case studies and real-world examples of AI in action helps students grasp the practical implications of these ethical, privacy, and societal issues (Russell & Norvig, 2016). It provides a context for abstract concepts, making the learning process more engaging and relatable.

5. **Critical Thinking and Decision-Making Skills**:

    ○ As AI systems increasingly make decisions previously made by humans, it's important for educational curricula to foster critical thinking and decision-making skills. This enables students to critically assess AI recommendations and understand the limitations of AI systems (Heaven, 2019).

6. **Interdisciplinary Approach**:

    ○ Teaching these topics requires an interdisciplinary approach, combining insights from computer science, law, philosophy, and social sciences. This approach ensures a well-rounded understanding of AI's multifaceted impact (Walsh, 2018).

7. **Encouraging Ethical Innovation**:

- Lastly, education should encourage ethical innovation, inspiring students to design AI solutions that are not only technologically advanced but also ethically sound and socially responsible (Bostrom & Yudkowsky, 2014).

VI. Personalized and Adaptive Learning Systems:

Leveraging AI in education itself, personalized learning systems can adapt to individual student's needs, pace, and learning style, making education more effective and inclusive (Xie et al., 2019).

The evolution of educational curricula in the AI age is significantly influenced by the development of personalized and adaptive learning systems. These AI-driven systems offer tailored educational experiences, catering to the individual needs, skills, and learning pace of each student.

1. **Personalized Learning Experiences**:
   - AI systems can analyze students' learning styles, strengths, and weaknesses to provide customized learning materials and activities (Xie et al., 2019). This individualized approach ensures that each student receives the support and challenges they need to succeed.

2. **Adaptive Learning Technologies**:
   - Adaptive learning technologies adjust the content and pace of learning based on real-time feedback from students' performance (Sottilare et al., 2018). This dynamic adjustment helps in maintaining an optimal level of challenge and engagement for each student.

3. **Data-Driven Insights**:
   - AI in education generates vast amounts of data on student learning patterns. Analyzing this data helps educators identify gaps in knowledge and skills, enabling them to tailor their teaching strategies effectively (Baker & Inventado, 2014).

4. **Continuous Assessment and Feedback**:
   - Personalized learning systems provide continuous assessment and instant feedback, allowing students to understand their progress and areas for improvement quickly (Kulik, 2016). This immediate feedback loop enhances learning efficiency and motivation.

5. **Accessible Education for Diverse Needs**:
   - AI-driven personalized learning systems can be particularly beneficial for students with special educational needs or those who require additional support, offering resources that cater to their unique learning requirements (Zhu et al., 2020).

6. **Scalability and Flexibility**:
   - These systems enable scalability in education, offering quality learning experiences to a larger number of students, irrespective of geographical or physical constraints (Li & Ma, 2020).

7. **Preparing for Future Challenges**:
   - By providing a personalized learning journey, these AI systems prepare students for future challenges, equipping them with the skills and knowledge required in an ever-changing world (Ferguson, 2019).

8. **Ethical and Privacy Considerations**:
   - As with all AI applications, ethical and privacy considerations are paramount. Protecting student data and ensuring it is used responsibly is a key concern in the deployment of personalized learning systems (Prinsloo & Slade, 2017).

VII. Preparing for a Lifelong Learning Mindset:

With the continuous evolution of AI, instilling a lifelong learning mindset is crucial. Educational systems should prepare students for ongoing learning and upskilling throughout their careers (Schwab & Samans, 2016).

In the era of rapid technological advancements, particularly with the advent of AI, the emphasis on evolving educational curricula must include preparing students for a lifelong learning mindset. This approach is critical for adapting to continuous changes and challenges in the job market and society.

1. **Cultivating a Culture of Continuous Learning**:
   - Educational institutions are increasingly focusing on instilling a culture of continuous learning among students. This involves teaching students to view education as an ongoing process rather than a finite goal (Smith, 2021). Encouraging curiosity and a love for learning is central to this approach.

2. **Skill Development beyond Academics**:
   - The curriculum needs to emphasize not just academic knowledge but also skills like adaptability, problem-solving, and critical thinking, which are crucial for lifelong learning (Trilling & Fadel, 2018). These skills enable individuals to continually adapt and grow in their professional and personal lives.

3. **Incorporation of AI and Technology in Learning**:
   - By integrating AI and emerging technologies into the learning process, students can be better prepared to utilize these tools for their lifelong educational journey (Luckin et al., 2016). Understanding how to leverage technology effectively is key in a world where new tools and platforms are constantly emerging.

4. **Promoting Self-Directed Learning**:
   - Encouraging self-directed learning helps students take control of their learning process. This includes setting their learning goals, finding resources, and assessing their progress (Knowles, 1975). Such an approach fosters independence and a proactive attitude towards learning.

5. **Resilience and Adaptability Training**:
   - ○ Preparing students for the challenges of the future workplace and society involves teaching them resilience and adaptability. This training helps them navigate through changes and uncertainties in their careers and life (Robertson, 2020).

6. **Collaborative Learning Environments**:
   - ○ Collaborative learning environments, both physical and virtual, facilitate the sharing of knowledge and experiences, fostering a community of learners who support each other's educational journeys (Laal & Laal, 2012).

7. **Lifelong Career Guidance**:
   - ○ Educational institutions are also looking at providing ongoing career guidance to alumni, helping them navigate career shifts and continuous professional development needs (Hillage & Pollard, 1998).

8. **Reflective Practices**:
   - ○ Encouraging reflective practices helps students assess their learning process, understand their progress, and identify areas for improvement, which is vital for lifelong learning (Boud et al., 1985).

VIII. Partnerships with Industry and Real-World Application:

Establishing partnerships with industry can help align educational curricula with real-world needs. Internships, apprenticeships, and collaboration on curriculum development ensure that students gain relevant skills (Hernández-de-Menéndez et al., 2019).

The rapidly changing landscape of the job market, particularly with the integration of AI and emerging technologies, necessitates a transformation in educational curricula. A significant aspect of this transformation is the development of partnerships with industry and a focus on real-world applications.

1. **Forming Strategic Partnerships with Industry Leaders**:
   - Educational institutions are increasingly collaborating with industry leaders to ensure that curricula are aligned with current and future job market requirements (Bridgstock, 2009). These partnerships facilitate the integration of practical skills and industry insights into academic programs.
2. **Curriculum Design Informed by Industry Needs**:
   - A key component of these partnerships is the active involvement of industry experts in curriculum development. This ensures that the skills and knowledge taught are relevant and up-to-date with industry standards and technological advancements (Oliver, 2015).
3. **Work-Integrated Learning Opportunities**:
   - Incorporating internships, co-op programs, and project-based learning into the curriculum provides students with hands-on experience in real-world settings. This exposure is crucial for understanding the practical applications of theoretical knowledge (Patrick et al., 2008).
4. **Focus on Practical Problem-Solving Skills**:
   - The curriculum should emphasize practical problem-solving skills, preparing students to tackle real-world challenges effectively. This approach fosters critical thinking and adaptability, essential in a rapidly evolving job market (Helyer, 2015).
5. **Guest Lectures and Workshops by Industry Professionals**:
   - Regular guest lectures and workshops conducted by industry professionals provide students with insights into current industry trends, challenges, and best practices (Jackson, 2013).
6. **Incorporation of Latest Technologies and Tools**:
   - Partnerships with industry allow educational institutions to access and incorporate the latest technologies and tools

into their curriculum, ensuring students are well-versed in current and emerging technologies (Perkins, 2014).

7. **Feedback Mechanisms for Continuous Improvement**:
   - Continuous feedback from industry partners helps institutions refine and update their curricula, keeping them relevant and effective (Eraut, 2004).

8. **Enhancing Employability Skills**:
   - By aligning curricula with industry needs, students are better prepared for the workforce, enhancing their employability skills and career prospects (Yorke, 2006).

IX. Global and Cultural Competence:

In a globally connected world, understanding different cultures and global issues is vital. AI education should include global perspectives and cultural competence to prepare students for international collaboration and understanding (Reimers & Chung, 2016).

In the context of a globalized and technologically advanced society, educational curricula must evolve to include a focus on global and cultural competence. This is essential for preparing students to navigate and contribute to an increasingly interconnected world.

1. **Integration of Global Awareness into Curriculum**:
   - Curricula should be designed to broaden students' understanding of diverse cultures and global issues. This includes studying global socio-economic trends, environmental challenges, and cultural practices, providing a holistic worldview (Mansilla & Jackson, 2011).

2. **Cultural Competence and Sensitivity Training**:
   - With the workforce becoming more diverse, it's imperative to include cultural competence training. This involves teaching students effective communication and collaboration skills across various cultures, enhancing their ability to work in multinational environments (Deardorff, 2006).

3. **Multilingual Education:**
   - Language learning is a key component of cultural competence. Offering multilingual education can equip students with the necessary tools to communicate effectively in a global context, breaking down barriers and fostering international collaboration (Byram, 2008).

4. **Technology-Enhanced Cross-Cultural Interactions:**
   - Leveraging AI and digital platforms to facilitate cross-cultural exchanges and collaborations can enhance students' global competence. Virtual exchange programs allow students to interact with peers from different parts of the world, gaining insights into diverse perspectives (O'Dowd, 2018).

5. **Incorporating Global Case Studies and Examples:**
   - Using case studies and examples from various cultures and countries in the curriculum helps students understand the global implications of local actions and decisions. It encourages them to think critically about their role in a global society (Green & Olson, 2008).

6. **Study Abroad and International Programs:**
   - Encouraging participation in study abroad programs or international internships can provide students with first-hand experience in different cultural settings. These experiences are invaluable in developing empathy, adaptability, and a global mindset (Twombly et al., 2012).

7. **Focus on Global Ethics and Responsibility:**
   - Educating students about global ethics and social responsibility, particularly in the context of AI and technology, prepares them to make ethical decisions that consider the broader impacts on society and the environment (Hanson, 2010).

8. **Emphasis on Collaborative Problem-Solving:**

○ Teaching students to work collaboratively on global challenges, such as climate change or health crises, fosters a sense of shared responsibility and global citizenship (Rios & Siveroni, 2012).

The importance of lifelong learning and adaptability:

In the rapidly evolving landscape of the AI age, the concept of lifelong learning and adaptability becomes crucial for both personal and professional growth. The continuous advancements in technology demand that individuals update their skills and knowledge throughout their careers.

1. **Necessity of Continuous Learning**:

With AI and automation changing the job market, the skills that are in demand today may not be relevant tomorrow. Lifelong learning ensures that individuals remain employable and competitive. This approach is supported by studies emphasizing the need for continuous skill development in the face of technological advancements (Brynjolfsson & McAfee, 2014).

In the age of rapid technological advancements, particularly with the emergence of AI, the necessity of continuous learning has become more pronounced than ever. This imperative stems from several key factors:

1. **Rapid Technological Changes**:
   ○ The fast pace of technological innovation, especially in AI, necessitates a corresponding acceleration in learning and skill acquisition. As AI and related technologies evolve, the skills required to work alongside these technologies also change (Bughin, Hazan, Ramaswamy, Chui, Allas, Dahlström, Henke, & Trench, 2018).

2. **Job Market Evolution:**

   - The job market is continuously evolving, with AI and automation transforming existing roles and creating new ones. This evolution demands that workers not only acquire new technical skills but also adapt their cognitive and social skills to remain relevant and employable (Manyika, Lund, Chui, Bughin, Woetzel, Batra, Ko, & Sanghvi, 2017).

3. **Shrinking Half-Life of Skills:**

   - The half-life of skills is decreasing, meaning the skills learned today may become obsolete more quickly than in the past. Continuous learning is crucial to keep pace with this rapid obsolescence of skills. The World Economic Forum (2018) highlights the shrinking half-life of professional skills, emphasizing the need for ongoing education and training.

4. **Enhancing Cognitive Flexibility:**

   - Continuous learning enhances cognitive flexibility, allowing individuals to adapt to new information and changing environments. This flexibility is essential in a world where AI and automation are constantly altering the way tasks are performed (Hase & Kenyon, 2007).

5. **Economic Imperative for Individuals and Nations:**

   - On an economic level, continuous learning is not just an individual necessity but also a national imperative. Economies that foster a culture of lifelong learning are better positioned to maintain competitiveness in the global market, particularly in sectors driven by AI and technology (Schwab, 2016).

6. **Fulfillment of Personal Potential:**

   - Beyond economic and employment considerations, continuous learning is crucial for personal growth and the fulfillment of human potential. Engaging in lifelong learning activities has been linked to increased personal satisfaction and well-being (Field, 2000).

7. **Preparation for Unpredictable Career Paths**:
   - Given the unpredictable nature of future career paths, where individuals may change roles or industries multiple times, continuous learning equips them with a diverse skill set and enhances their adaptability (Brynjolfsson & McAfee, 2014).

II. Adaptability in a Changing Work Environment:

Adaptability is key in an AI-driven world. As AI technologies evolve, workers must be able to quickly adapt to new tools, work practices, and changing job roles. This adaptability is crucial for career longevity and success (Schwab, 2016).

Adaptability in a rapidly changing work environment, especially in the context of the AI revolution, is a crucial skill for modern workers. This need for adaptability arises from several key factors:

1. **Technological Disruption**:
   - AI and automation are transforming traditional job roles, creating a dynamic work environment where adaptability is essential. Workers must be able to pivot and learn new skills as their roles evolve or as new roles emerge (Bughin, Hazan, Ramaswamy, Chui, Allas, Dahlström, Henke, & Trench, 2018).

2. **The Emergence of New Job Categories**:
   - As AI continues to grow, new job categories are emerging that require a combination of technical, cognitive, and social skills. Workers need to be adaptable to move into these new fields and capitalize on the opportunities they present (Manyika et al., 2017).

3. **The Need for Soft Skills**:

- In addition to technical skills, soft skills such as emotional intelligence, critical thinking, and adaptability itself are becoming increasingly important. These skills enable workers to effectively navigate a workplace where human-machine collaboration is commonplace (World Economic Forum, 2018).

4. **Continuous Learning Culture**:

- Organizations are fostering a culture of continuous learning to keep pace with technological changes. This culture requires employees to be adaptable, open to learning, and ready to apply new knowledge and skills in their work (Meister, 2020).

5. **Personal and Professional Growth**:

- Adaptability is not only a requirement for job survival but also a key driver of personal and professional growth. Adaptable individuals are more likely to embrace challenges, learn from them, and grow in their careers (Hase & Kenyon, 2007).

6. **Globalization and Cultural Agility**:

- The globalized nature of modern business requires workers to be culturally agile and adaptable to different working environments, practices, and team dynamics. This agility is particularly important in multinational teams and projects involving AI and technology (Schwab, 2016).

7. **Resilience in the Face of Change**:

- Adaptability contributes to resilience, enabling workers to cope with changes and uncertainties in the AI-driven workplace. Resilient workers are better equipped to handle the stress and challenges of rapid technological change (Robertson, Cooper, Sarkar, & Curran, 2015).

III. Personal Development and Growth:

Lifelong learning contributes to personal development by encouraging curiosity, creativity, and a growth mindset. Engaging in continuous learning can lead to greater personal satisfaction and improved mental health (Deci & Ryan, 2000).

Personal development and growth are increasingly recognized as critical components in lifelong learning and adaptability, particularly in the AI-driven age. This focus stems from several interconnected aspects:

1. **Enhancing Cognitive Flexibility**:
   - Lifelong learning fosters cognitive flexibility, allowing individuals to adapt to new information and changing environments. This adaptability is essential in a workplace continuously reshaped by AI and technology (Helsdingen & Van Gog, 2018).

2. **Building Emotional Intelligence**:
   - Emotional intelligence is a key factor in personal and professional success. Continuous learning and adaptability help individuals manage their emotions and understand others', a crucial skill in an AI-integrated work environment (Goleman, 1995).

3. **Promoting Mental Agility**:
   - Lifelong learning and adaptability contribute to mental agility, enabling individuals to think critically and creatively, solve problems efficiently, and adapt to new challenges — skills highly valued in the AI era (Dweck, 2006).

4. **Cultivating a Growth Mindset**:
   - A growth mindset, the belief that abilities can be developed through dedication and hard work, is fundamental for personal development. This mindset encourages continuous learning and adaptability, crucial in adapting to AI-driven changes (Dweck, 2007).

5. **Enhancing Career Prospects**:

- ○ Continuous learning and adaptability lead to higher employability and better career prospects. As AI transforms industries, individuals who invest in their personal development can navigate these changes more effectively (Schwab, 2016).

6. **Improving Life Satisfaction**:
   - ○ Lifelong learning has been linked to increased life satisfaction and well-being. Engaging in continuous learning and development activities can lead to a more fulfilling and balanced life (Jenkins & Mostafa, 2015).

7. **Fostering Resilience**:
   - ○ Adaptability and lifelong learning build resilience, enabling individuals to better handle stress and bounce back from challenges. This resilience is particularly important in the fast-paced, AI-driven work environment (Robertson et al., 2015).

8. **Encouraging Self-Reflection**:
   - ○ Lifelong learning involves continuous self-reflection and assessment, leading to greater self-awareness and personal growth. This reflective practice is essential for adapting to the evolving demands of the AI age (Boud, Keogh, & Walker, 1985).

IV. Economic Imperatives:

From an economic perspective, lifelong learning is essential for maintaining a competitive workforce. As economies become more knowledge-based and technologically driven, the ability of the workforce to learn and adapt is directly linked to economic growth and stability (OECD, 2019).

The importance of lifelong learning and adaptability in the context of economic imperatives, especially in the AI age, is multifaceted and crucial:

1. **Responding to Rapid Technological Changes**:
   - Lifelong learning is essential to keep pace with the rapid technological changes, particularly in AI and related fields. Workers must continuously update their skills to remain relevant and competitive in the job market (Autor, 2015).
2. **Fostering Innovation and Competitiveness**:
   - Continuous learning and adaptability are key drivers of innovation and competitiveness in a global economy. As AI transforms industries, workers who can adapt and innovate will contribute to economic growth (Bughin et al., 2018).
3. **Reducing Skills Gaps and Mismatches**:
   - Lifelong learning helps address the skills gap and mismatches in the labor market. As AI evolves, the demand for new skills emerges rapidly. Continuous learning ensures that the workforce can meet these evolving demands (World Economic Forum, 2018).
4. **Supporting Career Transitions and Mobility**:
   - In an AI-driven economy, many traditional roles are being redefined or becoming obsolete. Lifelong learning facilitates career transitions and enhances job mobility, crucial for economic stability and growth (OECD, 2019).
5. **Enhancing Productivity**:
   - Workers who engage in lifelong learning tend to be more productive. As AI and automation change the nature of work, a continuously learning workforce can better leverage these technologies to enhance productivity (Bessen, 2019).
6. **Promoting Employment Stability**:
   - Lifelong learning and adaptability contribute to employment stability. In an era where AI can disrupt job markets, those who are adaptable and committed to ongoing

learning are more likely to maintain stable employment (Acemoglu & Autor, 2011).

7. **Driving Economic Resilience**:
   - A workforce skilled in lifelong learning and adaptability is crucial for economic resilience, particularly in facing disruptions like those caused by AI advancements. Such a workforce can more effectively navigate and recover from economic shocks (ManpowerGroup, 2020).

8. **Encouraging Entrepreneurial Ventures**:
   - Lifelong learning fosters an entrepreneurial mindset, essential in the AI age. This mindset enables individuals to identify and capitalize on new opportunities created by AI, thus driving economic innovation and diversification (Shane, 2003).

V. The Role of Educational Institutions:

Educational institutions must adapt their curricula and teaching methodologies to foster lifelong learning. This includes integrating AI and technology into education and promoting self-directed learning strategies (Siemens, 2005).

The role of educational institutions in fostering lifelong learning and adaptability in the AI age is critical:

1. **Curriculum Development and Integration of AI**:
   - Educational institutions must revamp curricula to integrate AI and emerging technologies, ensuring that students are not only proficient in technical skills but also in understanding the implications of AI (Wagner et al., 2020).

2. **Promoting an Adaptive Mindset**:
   - Schools and universities play a pivotal role in cultivating an adaptive mindset. This involves teaching students to be flexible, curious, and resilient in the face of technological changes (Zheng & Walsh, 2019).

3. **Focus on Soft Skills**:
   - Besides technical knowledge, educational institutions should emphasize soft skills like critical thinking, problem-solving, and emotional intelligence, which are crucial in an AI-dominated workforce (Deming, 2017).

4. **Lifelong Learning Opportunities**:
   - Universities and colleges should offer continuing education and lifelong learning opportunities, allowing individuals to upskill or reskill at different stages of their careers (Goldin & Katz, 2008).

5. **Partnerships with Industry**:
   - Educational institutions should collaborate with industries to ensure that their programs are aligned with the current and future needs of the job market, particularly in AI-related fields (Bessen, 2015).

6. **Incorporating Experiential Learning**:
   - Hands-on, experiential learning opportunities, such as internships and project-based learning, are essential for students to apply theoretical knowledge in real-world contexts (Kolb, 2014).

7. **Access to Cutting-Edge Research**:
   - Universities, especially, have the responsibility to engage in cutting-edge research in AI and its societal impacts, thus informing and shaping their teaching and training programs (Frey & Osborne, 2017).

8. **Educational Equity and Access**:
   - Educational institutions must also address the digital divide by ensuring equitable access to AI education and training for all students, regardless of their background (Zhao, 2019).

9. **Ethics and Responsible AI**:
   - Teaching the ethical considerations and responsible use of AI is crucial. Institutions must embed these principles in

their curricula to prepare students for the moral challenges posed by AI technologies (Hagendorff, 2020).

10. **Global and Cultural Perspectives:**
    - As AI has a global impact, educational programs must include global and cultural perspectives, preparing students to work in diverse, international environments (Nussbaum, 2010).

VI. Impact on Society:

Lifelong learning plays a critical role in societal development. Educated and adaptable citizens are better equipped to contribute to their communities and address complex social challenges (Field, 2000).

The impact of lifelong learning and adaptability on society, especially in the context of the AI age, is profound:

1. **Economic Resilience:**
   - Lifelong learning and adaptability contribute to economic resilience by equipping the workforce to adapt to AI-induced job transformations, thus maintaining employment levels and economic stability (Schwab, 2016).

2. **Innovation and Productivity:**
   - A workforce that is continuously learning and adapting is more likely to drive innovation and productivity. This is particularly true in sectors heavily influenced by AI and technology (Brynjolfsson & McAfee, 2014).

3. **Social Inclusion and Equity:**
   - Lifelong learning opportunities help bridge the gap between various socio-economic groups, promoting social inclusion and equity. This is crucial in countering the inequalities often exacerbated by technological advancements (Castells, 2001).

4. **Health and Well-being:**

- Engaging in lifelong learning has been linked to improved health and well-being. The adaptability skills fostered through continuous learning can reduce stress and anxiety associated with AI and automation-related job disruptions (Vaillant & Mukamal, 2001).

5. **Civic Participation**:
   - Educated and adaptable citizens are more likely to engage in civic activities, contributing positively to democratic societies. Understanding and adapting to AI can also lead to more informed discussions and decisions about technology policies (Putnam, 2000).

6. **Cultural Enrichment**:
   - Lifelong learning, especially in a diverse set of subjects including those influenced by AI, contributes to cultural enrichment. It fosters an appreciation of different perspectives and enhances cultural awareness (Eisner, 2002).

7. **Global Competitiveness**:
   - On a global scale, a nation's commitment to lifelong learning and adaptability in its workforce enhances its competitiveness. Countries that invest in these areas are better positioned in the global economy, particularly in industries driven by AI (Porter & Stern, 2000).

8. **Environmental Sustainability**:
   - Lifelong learning and adaptability in the AI age can also contribute to environmental sustainability. Educated individuals are more likely to understand and advocate for sustainable practices, including those related to AI and technology use (Kates et al., 2001).

9. **Mitigating Ethical Risks of AI**:
   - As society adapts to AI, lifelong learning can help individuals understand and mitigate the ethical risks associated with AI, such as privacy concerns, bias, and job displacement (Bostrom, 2014).

10. **Cultural Integration of AI**:
    - Lifelong learning helps society integrate AI technologies in a way that is culturally sensitive and informed, thereby ensuring that AI advancements are in harmony with societal values and norms (Susskind & Susskind, 2015).

VII. Personalized Learning Paths:

The advancement of AI in education allows for personalized learning paths, making lifelong learning more accessible and effective. AI can help identify individual learning needs and provide tailored resources (Hwang, 2004).

The concept of personalized learning paths in the context of lifelong learning and adaptability is becoming increasingly significant in the AI age:

1. **Tailored Education**:
    - Personalized learning paths allow education to be tailored to individual needs, strengths, and interests, enhancing the learning experience and outcomes (Pane et al., 2017). This individualization is particularly important in adapting to diverse career paths influenced by AI.

2. **Technology Integration**:
    - AI and machine learning technologies enable the creation of adaptive learning systems that can modify content and pace based on learner performance and preferences (Baker & Smith, 2019). This integration is crucial for effective lifelong learning in a technology-driven world.

3. **Enhanced Engagement and Motivation**:
    - Personalized learning paths have been shown to increase student engagement and motivation, as learners feel more connected to the content that is relevant to their interests and career goals (Deci & Ryan, 2000).

4. **Continuous Skill Development**:

- ○ In the rapidly evolving job market, personalized learning paths facilitate continuous skill development, ensuring individuals remain competitive and adaptable (Fosway Group, 2018).

5. **Bridging Skill Gaps**:

- ○ These paths can effectively bridge skill gaps, particularly in areas impacted by AI and automation, by focusing on specific skills and competencies needed in the modern workforce (World Economic Forum, 2020).

6. **Access to Diverse Resources**:

- ○ Personalized learning embraces a variety of learning resources and formats, catering to different learning styles and preferences, which is essential in a diverse and inclusive learning environment (Honey & Hilton, 2011).

7. **Feedback and Assessment**:

- ○ Real-time feedback and assessment, integral to personalized learning paths, enable learners to recognize their progress and areas for improvement, fostering a growth mindset crucial for lifelong learning (Dweck, 2006).

8. **Career Advancement**:

- ○ These learning paths support career advancement by aligning learning objectives with career goals, making education more relevant and beneficial for professional growth (Boud & Garrick, 1999).

9. **Inclusion and Equity**:

- ○ Personalized learning can contribute to educational equity by providing learners from diverse backgrounds with resources and support tailored to their unique circumstances (Darling-Hammond et al., 2020).

10. **Fostering Lifelong Learning**:

- ○ By providing engaging, relevant, and adaptive learning experiences, personalized learning paths cultivate a lifelong learning mindset, an essential attribute in the AI age (Knowles, Holton, & Swanson, 2005).

VIII. Preparation for an Uncertain Future:

In a world where the future of work is uncertain, lifelong learning provides individuals with a sense of control and preparedness. It equips them with the skills to navigate and adapt to unforeseen changes (Frey & Osborne, 2017).

The AI age underscores the necessity of preparing for an uncertain future through lifelong learning and adaptability:

1. **Rapid Technological Advancements**:
   - The pace of technological change, especially in AI, requires individuals to continually update their skills to stay relevant. Continuous learning is essential to keep up with these advancements (Schwab, 2016).

2. **Job Market Fluidity**:
   - AI and automation are transforming job markets, creating new roles while rendering others obsolete. Lifelong learning enables individuals to adapt to these shifts, ensuring ongoing employability (Manyika et al., 2017).

3. **Uncertainty and Complexity**:
   - The future work environment is characterized by uncertainty and complexity, necessitating a workforce that is flexible and adaptable to unforeseen changes (Hagel et al., 2016).

4. **Resilience in the Face of Change**:
   - Lifelong learning fosters resilience, allowing individuals to better manage and adapt to change, a crucial skill in an unpredictable future (Robertson et al., 2015).

5. **Critical Thinking and Problem-Solving**:
   - As AI handles more routine tasks, the human workforce needs to focus on critical thinking and problem-solving,

skills that require continuous learning and adaptation (Davies et al., 2011).

6. **Interdisciplinary Knowledge**:
   - The future will demand interdisciplinary knowledge, as the boundaries between different fields blur due to technological integration. Lifelong learning supports the acquisition of diverse skills and knowledge (Carnegie Mellon University, 2018).

7. **Innovation and Creativity**:
   - Lifelong learning nurtures innovation and creativity, key competencies in an AI-driven world where the ability to innovate is a significant differentiator (Florida, 2002).

8. **Global Perspective**:
   - Understanding global trends and cultural diversity is vital in a connected world. Lifelong learning helps individuals gain a global perspective, crucial for navigating the international aspects of AI and technology (Friedman, 2005).

9. **Ethical Considerations**:
   - As AI evolves, ethical considerations become more complex. Continuous education in ethics is necessary to navigate these challenges responsibly (Wallach & Allen, 2009).

10. **Personal Fulfillment**:
    - Beyond professional needs, lifelong learning contributes to personal fulfillment and well-being, helping individuals find meaning and purpose in a rapidly changing world (Csikszentmihalyi, 1990).

Case studies: Successful retraining and upskilling programs:

In the age of AI, retraining and upskilling programs play a critical role in workforce development. Several case studies illustrate successful implementation:

1. **Amazon's Upskilling 2025 Program**:

Amazon's Upskilling 2025 Program serves as a leading example of how companies can proactively prepare their workforce for the evolving demands of the AI age. This initiative reflects a significant corporate investment in employee development and skill adaptation.

1. **Overview of the Program**:
    - Launched in 2019, Amazon committed $700 million to upskill 100,000 U.S. employees by 2025. This program aims to provide employees with the skills necessary for in-demand jobs within Amazon, emphasizing the areas most influenced by AI and automation technologies (Amazon, 2019).
2. **Key Components of the Program**:
    - Amazon's approach includes several specific training programs like the Amazon Technical Academy, Machine Learning University, and AWS Training and Certification. These programs are tailored to empower non-technical employees to transition into software engineering roles and technical employees to advance their skills in machine learning and cloud computing (Amazon, 2019).
3. **Program's Impact and Success**:
    - The Upskilling 2025 Program has shown success in enabling employees to transition to higher-paying, tech-focused jobs within Amazon. This initiative not only benefits the employees by enhancing their career prospects but also helps Amazon build a more skilled and adaptable workforce. It's a model demonstrating the potential for large-scale corporate retraining initiatives in response to technological advancements (Amazon Annual Report, 2020).
4. **Future Implications**:

○ Amazon's program highlights the role of private corporations in workforce development in the AI age. It sets a precedent for other companies to invest in upskilling initiatives, recognizing the mutual benefits for employees and the organization. The program underscores the importance of continuous learning and adaptability in a technology-driven job market (Bersin, 2019).

5. **Critiques and Challenges**:

○ While widely praised, the program has also faced critiques. Some argue that it primarily serves Amazon's interests, addressing the company's need for tech talent but not necessarily leading to broader societal benefits. Additionally, the challenge of scaling such programs to smaller companies without Amazon's resources remains a concern in the broader context of workforce development (Kaplan, 2020).

II. **AT&T's Future Ready Program**:

AT&T's Future Ready Program is a prime example of a corporate initiative designed to prepare its workforce for the digital era, particularly in the context of rapid advancements in AI and technology.

1. **Program Overview**:

○ Initiated in 2016, AT&T's Future Ready Program is an extensive effort to retrain and upskill its existing workforce. The program was designed in response to the growing need for employees skilled in areas like data science, cybersecurity, and network management, which are critical in the age of AI and digital transformation (AT&T, 2016).

2. **Key Elements of the Program**:

○ AT&T's approach includes online courses, collaborative learning platforms, and professional development opportunities. The program offers a mix of self-paced learning and structured curricula, with a focus on emerging technologies and digital skillsets. Partnerships with online educational platforms like Coursera and Udacity play a significant role in providing accessible learning resources to employees (Stephenson, 2016).

3. **Impact and Results**:

   - The Future Ready Program has enabled a significant portion of AT&T's workforce to adapt to new roles and responsibilities. By 2020, over 70% of employees had engaged in new learning opportunities, leading to better preparedness for the evolving digital landscape. This investment in human capital has not only enhanced individual careers but also ensured that AT&T remains competitive in the technology sector (AT&T Annual Report, 2020).

4. **Long-term Implications**:

   - This initiative illustrates the critical role of continuous learning in corporate settings, especially as AI and digital technologies continue to evolve. AT&T's program serves as a model for other companies facing similar transitions, highlighting the importance of investing in employee training and development for future readiness (Meister, 2017).

5. **Challenges and Considerations**:

   - Despite its success, the program faces challenges, such as ensuring ongoing engagement and balancing the immediate operational needs with long-term learning goals. Furthermore, adapting such a program to smaller companies or different industries remains a complex task, given the significant resources required for such large-scale training initiatives (Boudreau & Ziskin, 2019).

III. **Singapore's SkillsFuture Initiative**:

The SkillsFuture Initiative in Singapore is a comprehensive national program designed to foster a culture of lifelong learning and equip Singaporeans with skills for the future, especially in the face of AI and technological advancements.

1. **Program Overview**:
   - Launched in 2015, the SkillsFuture Initiative is a government-led strategy aimed at enabling Singaporeans to develop necessary skills throughout their lives, with a particular focus on skills relevant in the AI and digital era. The program supports a wide range of courses and learning opportunities across various sectors (SkillsFuture Singapore, 2015).

2. **Key Elements of the Program**:
   - SkillsFuture encompasses various components, including subsidies for courses related to emerging technologies, career guidance, and a credit system for Singaporeans to fund their learning endeavors. The program collaborates with educational institutions, training providers, and industries to ensure a broad and relevant array of learning opportunities (Ministry of Education, Singapore, 2016).

3. **Impact and Results**:
   - Since its inception, SkillsFuture has seen wide participation, with many Singaporeans utilizing their credits to enroll in courses that enhance digital literacy, AI understanding, and other future-ready skills. The initiative has been pivotal in fostering a proactive approach to upskilling among the Singaporean workforce (Ng, 2018).

4. **Long-term Implications**:
   - The SkillsFuture Initiative demonstrates the effective role of national strategies in workforce development for the AI age. It highlights how government policies and support can facilitate continuous learning and adaptability among

citizens, which is crucial in a rapidly evolving job market (Tan, 2019).

5. **Challenges and Considerations**:
   - One of the challenges faced by the initiative is ensuring that the offered skills training remains aligned with the fast-changing demands of the labor market. Additionally, engaging various demographics, especially older workers, in lifelong learning presents ongoing challenges (Lee & Zhao, 2020).

IV. Google's IT Support Professional Certificate:

Google's IT Support Professional Certificate is a notable example of a successful retraining and upskilling program, specifically designed to address the growing need for IT support professionals in the era of AI and digital transformation.

1. **Program Overview**:
   - Launched in 2018, this program is hosted on Coursera and aims to prepare individuals for entry-level roles in IT support in about six months. It is part of Grow with Google, an initiative to help create economic opportunities for all (Grow with Google, 2018).

2. **Curriculum and Learning Objectives**:
   - The program covers fundamental IT support topics such as computer assembly, networking, system administration, and security, with a significant emphasis on real-world applications. It also includes problem-solving and customer service, skills crucial in IT support roles (Coursera, 2018).

3. **Accessibility and Inclusivity**:
   - A key feature of this program is its accessibility. It requires no previous experience and offers financial aid, making it inclusive for people from diverse backgrounds, including those making career transitions (Google, 2019).

4. **Impact and Results**:
   - Google reported that 82% of graduates from the IT Support Professional Certificate program in the U.S. reported a positive career outcome like a new job, enhanced skills, or promotion within six months of completion (Google, 2020).

5. **Partnerships and Industry Integration**:
   - Google has partnered with over 100 companies like Walmart, Hulu, and Sprint, to consider graduates of the program for relevant IT roles. This industry integration is pivotal for practical applicability and employment prospects (Google, 2020).

6. **Long-term Implications**:
   - This initiative showcases the potential of corporate-led education programs in bridging skill gaps in technology-driven fields. It aligns with the need for continuous learning and skill development in the AI age, demonstrating the role of private sector initiatives in workforce development (Smith, 2021).

7. **Challenges and Considerations**:
   - One challenge is keeping the curriculum aligned with the rapidly evolving IT landscape. Another consideration is ensuring the program remains accessible and relevant to a global audience, considering different regions' unique IT landscapes (Johnson, 2022).

V. IBM's New Collar Jobs Initiative:

IBM's New Collar Jobs Initiative represents a significant stride in retraining and upskilling programs, particularly in addressing the skills gap in the technology sector amidst the rise of AI and digital innovation.

1. **Initiative Overview**:

○ Launched in 2017, the New Collar Jobs Initiative focuses on roles in technology sectors that don't necessarily require a traditional college degree. IBM emphasizes skills and practical experience over formal education for these positions (Rometty, 2017).

2. **Targeted Skills and Roles**:

○ The initiative targets roles in areas like cybersecurity, cloud computing, and digital design, which are critical in the current technology landscape. It seeks to equip participants with specific, in-demand skills rather than broad-based academic qualifications (IBM, 2018).

3. **Training Programs and Methodology**:

○ IBM's approach includes vocational training, coding camps, professional certification programs, and innovative educational models like P-TECH (Pathways in Technology Early College High School), which combines high school and community college education with career experience (IBM, 2019).

4. **Partnerships and Collaboration**:

○ A key aspect is collaboration with educational institutions and governments. IBM partners with high schools, colleges, and other entities to build relevant curriculum and training pathways that lead directly to employment opportunities (P-TECH, 2019).

5. **Impact and Success Metrics**:

○ IBM reports significant success in filling positions through this initiative, with thousands of new collar job hires. It serves as a model for how companies can adapt their workforce strategies to the evolving demands of the AI age (IBM Annual Report, 2020).

6. **Challenges and Future Prospects**:

○ The challenge lies in scaling and adapting these programs to various global regions with different educational and economic backgrounds. The future prospect involves expanding the initiative to more fields and geographies (Brown & Lauder, 2021).

7. **Policy and Societal Implications**:

   ○ This initiative highlights the importance of policy support for non-traditional education pathways and the need for public-private partnerships in workforce development. It underscores the shift in job requirements in the digital era (Schwartz et al., 2022).

VI. **P-TECH Schools**:

Pathways in Technology Early College High School (P-TECH) is a pioneering educational model that blends high school, college, and industry experience, focusing on equipping students with skills relevant in the AI age.

1. **Program Overview**:

   ○ Initiated in 2011 through a collaboration between IBM, the New York City Department of Education, and the City University of New York, P-TECH aims to create a seamless pathway from high school to college and into the professional world (Rometty, 2016).

2. **Educational Model**:

   ○ P-TECH schools offer a six-year program where students earn a high school diploma and an associate degree in a STEM field. The curriculum is designed in partnership with industry leaders to ensure alignment with job market needs, particularly in areas impacted by AI and digital technologies (Litow, 2018).

3. **Industry Partnerships**:

   ○ A critical component is the partnership with companies like IBM, which provide mentorship, internships, and potential job opportunities post-graduation. These partnerships help bridge the gap between education and the evolving needs of the workforce (IBM, 2019).

4. **Accessibility and Diversity:**

   ○ P-TECH schools prioritize accessibility, targeting under-served communities and aiming to increase diversity in tech-related fields. This approach helps address equity issues in education and employment (Davis, 2020).

5. **Global Expansion and Impact:**

   ○ Since its inception, P-TECH has expanded globally, with schools in multiple countries. Its success is evident in the high graduation rates and job placements, showcasing the effectiveness of this model in preparing students for future job markets (P-TECH, 2021).

6. **Challenges and Adaptations:**

   ○ Despite its success, P-TECH faces challenges like scaling the model and adapting to different educational systems and cultural contexts. Ongoing adaptations are required to keep the curriculum relevant as the AI and tech landscapes evolve (Johnson, 2022).

7. **Policy and Educational Implications:**

   ○ P-TECH's model highlights the importance of innovative educational structures and public-private partnerships in addressing the skills gap. It underscores the need for educational policies that support such collaborative models (Schwartz & Pellegrino, 2021).

VII. Salesforce's Trailhead Program:

Salesforce's Trailhead program is a standout example of a successful retraining and upskilling initiative in the age of AI and digital transformation.

1. **Program Overview:**

○ Launched in 2014, Trailhead is Salesforce's free online learning platform designed to teach people the skills needed to use and develop for Salesforce products. It's notable for its accessibility and flexibility, allowing learners to acquire skills at their own pace and on their own terms (Salesforce, 2014).

2. **Learning Approach**:

○ Trailhead adopts a gamified, self-paced learning approach. Users progress through various modules and projects, earning badges and points, which makes learning engaging and measurable. The content covers a wide range of topics, from basic CRM principles to advanced AI and data analytics (Miller, 2017).

3. **Focus on AI and Emerging Technologies**:

○ The program incorporates modules specifically focused on AI and emerging technologies like Salesforce's AI tool, Einstein. This ensures that learners are not only proficient in current technologies but are also prepared for future advancements (Bersin, 2019).

4. **Career and Community Impact**:

○ Trailhead has a significant career impact. Learners who complete the program can showcase their badges on their LinkedIn profiles, enhancing their employability. Salesforce also has a community-driven approach, with numerous Trailhead groups and events facilitating networking and collaborative learning (Salesforce, 2021).

5. **Global Reach and Diversity**:

○ Trailhead's global reach and commitment to diversity are evident in its multilingual offerings and focus on providing opportunities for underrepresented groups in tech. Salesforce actively works to ensure that its platform is inclusive and accessible to a diverse global audience (Gupta, 2020).

6. **Industry Recognition and Certifications**:

- ○ Trailhead also offers professional certifications, recognized in the industry, which can be a stepping stone for individuals seeking career advancement or shifts. This certification aspect adds tangible value to the skills learned on the platform (Smith, 2018).

7. **Impact on Lifelong Learning Culture**:

- ○ The Trailhead program contributes to fostering a culture of lifelong learning. By providing ongoing updates and new modules, it encourages continuous skill development, aligning with the dynamic nature of the tech industry (Brown, 2022).

VIII. **Walmart's Live Better U Program**:

Walmart's Live Better U (LBU) program is an illustrative case study in the realm of corporate-led retraining and upskilling programs, particularly in the age of AI and rapid technological change.

1. **Program Overview**:

- ○ Walmart introduced the Live Better U program in 2018 as a comprehensive education benefit for its employees. The program covers the cost of tuition, books, and fees, allowing Walmart associates to earn degrees or certificates in various fields without the burden of educational expenses (Walmart, 2018).

2. **Educational Partnerships**:

- ○ A key aspect of LBU is its partnerships with accredited universities and learning institutions. These partnerships enable employees to access a variety of learning paths, including those related to AI, data science, and technology, which are crucial in today's digital economy (Fernandez, 2019).

3. **Focus on Digital Upskilling**:

- Recognizing the importance of digital competence in the retail industry, particularly with the integration of AI and automation, LBU includes courses and modules focused on digital skills and technologies. This ensures that employees are not only competent in their current roles but are also prepared for future technological advancements (Johnson, 2020).

4. **Accessibility and Inclusivity**:

- The LBU program is designed to be accessible to a diverse workforce. By offering online and flexible learning options, Walmart accommodates the varied schedules and commitments of its employees, making education more inclusive and equitable (Baker, 2021).

5. **Impact on Employee Development and Retention**:

- The LBU program has had a significant impact on employee development and retention. By investing in employees' education, Walmart has seen improvements in job satisfaction and loyalty, as well as a better-prepared workforce to take on new challenges in the evolving retail landscape (Walmart, 2020).

6. **Strategic Alignment with Business Goals**:

- Walmart's commitment to this program aligns with its broader business strategy of integrating AI and technology into its operations. By upskilling its workforce, Walmart ensures that its employees are capable of working alongside emerging technologies, thus maintaining its competitive edge in the retail sector (Smith, 2021).

7. **Societal Impact and Corporate Responsibility**:

- The program also reflects Walmart's recognition of its corporate responsibility in contributing to societal advancement. By providing educational opportunities, Walmart plays a role in addressing the broader skills gap in the economy, particularly in tech-related fields (Garcia, 2022).

IX. **Finland's AI Education Program for Citizens**:

Finland's AI Education Program for Citizens represents a unique and forward-thinking approach to national skill development in the age of artificial intelligence.

1. **Program Overview**:
   - Launched by the Finnish government, this program aims to educate 1% of the Finnish population in the basics of AI. The initiative, known as "Elements of AI," is a free online course designed to demystify AI and empower citizens with knowledge about how AI works and its implications (Finnish Government, 2019).

2. **Accessibility and Inclusivity**:
   - The course is designed for individuals with no prior knowledge of AI, making it accessible to a broad audience. This inclusivity aligns with Finland's goal of enhancing digital literacy across its entire population, regardless of their professional background or age (Koivisto, 2020).

3. **Partnership with Industry and Academia**:
   - The program was developed in collaboration with the University of Helsinki and technology companies. This partnership ensures that the content is both academically sound and relevant to current industry practices (University of Helsinki, 2019).

4. **Empowering Citizens in an AI World**:
   - By educating citizens about AI, the program aims to reduce fear and misconceptions about AI technologies, fostering a more informed and engaged populace that is better prepared to interact with AI systems both in the workplace and in daily life (Niemi, 2021).

5. **Economic and Societal Impact**:
   - The Finnish AI education program is not just about individual empowerment; it's also seen as a strategic

investment in Finland's future. By building a population versed in AI, Finland is positioning itself as a leader in the digital economy (Helsinki Times, 2019).

6. **Scaling and International Adoption**:
    - The success of the program in Finland has led to its adaptation and adoption in other countries, demonstrating the scalability of such an initiative. This global expansion underscores the universal relevance of AI education in today's interconnected world (European Commission, 2020).

7. **Long-Term Vision for a Technologically Advanced Society**:
    - Finland's approach reflects a long-term vision to create a society that is not only equipped to use AI but also to contribute to its ethical and responsible development. This is in line with Finland's commitment to technological advancement while ensuring societal well-being (Korhonen, 2021).

X. Ericsson's Global AI Accelerator Program:

Ericsson's Global AI Accelerator Program is a significant initiative in the realm of corporate-led AI education and skill development.

1. **Program Goals and Structure**:
    - The Global AI Accelerator Program is designed to foster innovation and expertise in AI and machine learning (ML) within Ericsson. It focuses on accelerating the implementation of AI-driven solutions across various company operations (Ericsson, 2020).

2. **Employee Upskilling and Innovation**:
    - A key objective of the program is to upskill Ericsson's workforce in AI and ML. This involves providing employees with training sessions, workshops, and hands-on

projects that enhance their understanding and practical skills in these technologies (Johansson, 2021).

3. **Collaboration with Educational Institutions**:
   - Ericsson collaborates with leading universities and research institutes to ensure that the content and training provided are cutting-edge and relevant. This collaboration helps bridge the gap between academic research and industry application (Bergman, 2020).

4. **Global Reach and Diversity**:
   - The program has a global scope, involving participants from various regions and backgrounds. This diversity enriches the learning experience and fosters a more comprehensive understanding of AI applications in different market contexts (Ericsson Global, 2019).

5. **Impact on Business and Innovation**:
   - By investing in AI skills, Ericsson is enhancing its competitive edge in telecommunications and technology. The program not only benefits the employees but also contributes to the company's overall strategy in innovation and technology leadership (Nilsson, 2022).

6. **Focus on Ethical AI Development**:
   - The program emphasizes the importance of ethical considerations in AI development. It includes modules on responsible AI, ensuring that employees are aware of the ethical implications and societal impacts of AI technologies (Ericsson Ethics Board, 2021).

7. **Evaluation and Continuous Improvement**:
   - Ericsson continuously evaluates the effectiveness of the training programs, gathering feedback to improve future iterations. This approach ensures that the program remains relevant and effective in keeping pace with the rapidly evolving AI landscape (Ericsson HR, 2022).

8. **Contribution to Industry Standards in AI**:

○ Through its Global AI Accelerator Program, Ericsson is contributing to setting industry standards in AI and ML. The company shares insights and best practices, influencing the broader discourse on AI skill development in the tech industry (International Telecommunication Union, 2021).

12

Chapter 7: Economic Implications of an AI Workforce

The integration of Artificial Intelligence (AI) into the workforce is fundamentally transforming the economic landscape. Its implications are multifaceted, affecting labor markets, productivity, and the very nature of work itself.

1. **Job Displacement and Creation**:

AI has the potential to automate a range of tasks, leading to the displacement of certain jobs, particularly those involving routine or repetitive tasks (Frey & Osborne, 2017). However, it also creates new job opportunities, particularly in AI development, data analysis, and system maintenance (Smith & Anderson, 2014).

The arrival of Artificial Intelligence (AI) in the workforce has profound implications for job displacement and creation, reshaping the labor market in significant ways.

1. **Job Displacement**:
   - AI's capacity to automate tasks, especially those that are repetitive or routine, poses a risk of job displacement. Studies by Frey and Osborne (2017) indicate that up to 47% of US jobs are at high risk of automation. This displacement is more likely to affect lower-skilled jobs or roles that do not require complex decision-making or emotional intelligence (Acemoglu & Restrepo, 2018).

2. **Job Creation**:
   - Contrary to the fears of widespread job losses, AI also creates new job opportunities. The development, implementation, and maintenance of AI systems require skilled workers. Roles like AI specialists, data scientists, and AI ethics compliance officers are emerging (World Economic Forum, 2020).

3. **Transformation of Existing Jobs**:
   - Many existing jobs will not be eliminated but transformed. AI can augment human skills in jobs, leading to a change in job descriptions and required competencies. The World Economic Forum (2020) predicts a shift towards more strategic and creative roles as AI handles more routine tasks.

4. **Skills Gap and Reskilling**:
   - The displacement and creation of jobs due to AI highlight the importance of reskilling. The McKinsey Global Institute (2017) emphasizes the urgent need for large-scale upskilling initiatives to prepare the workforce for the future job market.

5. **Sector-Specific Impacts**:
   - The impact of AI on jobs varies by sector. Sectors like manufacturing and logistics may see higher displacement due to automation, while others like healthcare and

education may experience job growth due to AI's potential to enhance services (Autor, 2015).

6. **Geographical Variations**:

   ○ The impact of AI on jobs is not uniform across regions. Areas with economies focused on industries susceptible to automation may face greater challenges, while those with a high concentration of tech industries might see job growth (Manyika, Chui, Miremadi, Bughin, George, Willmott, & Dewhurst, 2017).

7. **Policy Implications**:

   ○ Governments and policymakers need to address the challenges posed by AI-induced job displacement. This includes investing in education, facilitating workforce transitions, and developing social safety nets for those affected by job losses (Susskind & Susskind, 2015).

II. Productivity and Efficiency:

AI can significantly enhance productivity and efficiency. By automating processes and providing insights through data analysis, AI enables businesses to operate more effectively (Bughin, Hazan, Ramaswamy, Chui, Allas, Dahlström, Henke, & Trench, 2018).

The integration of Artificial Intelligence (AI) in the workforce significantly enhances productivity and efficiency, leading to notable economic implications.

1. **Increased Productivity**:

   ○ AI technologies have the potential to vastly improve productivity. According to a report by Accenture (2016), AI could double annual economic growth rates by 2035 through changing the nature of work and creating new relationships between man and machine.

2. **Efficiency Gains**:

- AI-driven automation enables businesses to operate more efficiently. Tasks that are repetitive or require data processing can be completed quicker and with fewer errors, resulting in cost savings and increased output (Brynjolfsson & McAfee, 2017).

3. **Impact on GDP**:

- The efficiency and productivity gains from AI could lead to significant increases in GDP. PwC (2017) predicts that global GDP could be up to 14% higher in 2030 as a result of AI - the equivalent of an additional $15.7 trillion.

4. **Sector-Specific Productivity**:

- The impact of AI on productivity varies across sectors. Industries like manufacturing, logistics, and customer service may see considerable productivity boosts due to automation and optimization (Bughin, Hazan, Ramaswamy, Chui, Allas, Dahlström, Henke, & Trench, 2018).

5. **Redistribution of Labor**:

- While AI increases productivity in some areas, it also necessitates the redistribution of labor. As certain tasks become automated, workers may need to shift to roles that require more complex and creative skills (Autor, 2015).

6. **Cost Reduction and Scaling**:

- AI can significantly reduce operational costs for businesses by automating high-volume tasks and enabling scale without proportional increases in labor costs (Agrawal, Gans, & Goldfarb, 2018).

7. **Innovation and New Business Models**:

- The efficiency and capabilities provided by AI open up opportunities for new business models and innovation, leading to the development of new markets and industries (Bughin et al., 2018).

8. **Challenges in Measurement**:

- Quantifying the productivity gains from AI can be challenging. Traditional metrics may not fully capture the value created by AI, especially in service-oriented sectors (Syverson, 2017).

III. Economic Growth:

The adoption of AI is expected to contribute to economic growth. Research by PwC (2017) suggests that AI could contribute up to $15.7 trillion to the global economy by 2030, boosting global GDP by 14%.

The introduction of Artificial Intelligence (AI) into the workforce is a transformative factor that influences economic growth in various ways.

1. **Stimulating Economic Expansion**:
   - AI technologies have the potential to significantly accelerate economic growth. A study by PwC (2017) projected that AI could contribute up to $15.7 trillion to the global economy by 2030, which would increase global GDP by 14%.
2. **Driving Productivity**:
   - The efficiency and productivity gains from AI applications contribute to economic growth. According to a report by Accenture (2016), AI could potentially double the annual economic growth rates in some economies by 2035.
3. **Sectoral Transformation**:
   - Different sectors, such as manufacturing, healthcare, and finance, are experiencing a transformation due to AI, leading to new growth opportunities (Bughin et al., 2018).
4. **Innovation and New Markets**:
   - AI fosters innovation, creating new markets and products, thereby expanding economic activity (Brynjolfsson & McAfee, 2017).
5. **Impact on Employment and Wages**:

- ○ While AI can displace certain jobs, it also creates new ones, potentially leading to higher wages and new skill demands in the economy (Autor, 2015).

6. **Enhancing Global Trade**:

- ○ AI can optimize supply chains and logistics, enhancing global trade efficiency and thereby contributing to economic growth (Manyika et al., 2017).

7. **Regional Economic Disparities**:

- ○ The impact of AI on economic growth may vary regionally, potentially widening disparities between areas with different levels of AI adoption (Agrawal, Gans, & Goldfarb, 2018).

8. **Long-Term Structural Changes**:

- ○ The adoption of AI may lead to structural changes in the economy, with long-term implications for economic growth patterns (Acemoglu & Restrepo, 2018).

9. **Investment in AI and R&D**:

- ○ Increased investment in AI research and development can spur economic growth, as new technologies often lead to increased productivity and new economic opportunities (Aghion, Jones, & Jones, 2017).

10. **Challenges in Policy and Regulation**:

- ○ Policymakers face challenges in adapting to the rapid pace of AI development, ensuring that economic benefits are maximized while addressing potential negative impacts (West, 2018).

IV. Inequality and Wage Polarization:

The rise of AI could exacerbate income inequality. High-skilled workers who can leverage AI are likely to see wage increases, while low-skilled workers may face wage stagnation or job loss (Brynjolfsson & McAfee, 2014).

The integration of Artificial Intelligence (AI) into the workforce has significant implications for inequality and wage polarization. This topic explores various facets of how AI is reshaping the economic landscape in terms of distributional outcomes.

1. **Exacerbating Income Inequality**:
   - AI and automation can exacerbate income inequality. As Frey and Osborne (2017) point out, these technologies tend to replace low-skill jobs, widening the income gap between high-skill and low-skill workers.
2. **Wage Polarization**:
   - The deployment of AI leads to a polarization of wages. Autor, Levy, and Murnane (2003) demonstrate that automation can lead to a decline in middle-skill jobs, causing a polarization between high-wage and low-wage occupations.
3. **Skill Premium**:
   - AI increases the demand for high-skilled workers, resulting in a skill premium. Acemoglu and Autor (2012) discuss how technological advancements have historically favored skilled labor, leading to increased wage disparities.
4. **Displacement of Routine Jobs**:
   - AI's capability to automate routine tasks disproportionately affects certain job sectors, potentially displacing workers in these areas. This is highlighted in a study by Brynjolfsson and McAfee (2014), who note the vulnerability of routine jobs to AI and automation.
5. **Geographical Disparities**:
   - The impact of AI on employment and wages can vary significantly across different regions. A report by the McKinsey Global Institute (2019) notes that areas with a higher concentration of industries susceptible to automation face greater risks of inequality.

6. **Impact on Low-Income Workers**:
   - Low-income workers are particularly vulnerable to job displacement by AI, potentially exacerbating existing economic disparities. This is highlighted in the research by Chetty et al. (2016), examining the impact of technological change on different income groups.

7. **Role of Education and Training**:
   - The widening wage gap can be mitigated through education and retraining programs. A study by Goldin and Katz (2008) emphasizes the importance of education in equipping workers with the skills needed in an AI-driven economy.

8. **Policy Interventions**:
   - To address wage polarization and inequality, policymakers need to consider interventions such as income support, retraining programs, and changes in taxation, as discussed by Saez and Zucman (2019).

9. **Long-Term Economic Structure Changes**:
   - The long-term impact of AI on the economic structure could result in a fundamental shift in the nature of work and compensation, as explored by Aghion, Jones, and Jones (2019).

V. Changing Skill Demands:

The AI-driven economy demands a shift in skills. There is an increasing need for digital literacy, problem-solving, and continuous learning to adapt to new technologies (Schwab, 2016).

The advent of AI in the workforce has significantly altered skill demands, impacting both the nature of work and the skills required for future jobs. This change has various economic implications.

1. **Shift from Routine to Cognitive Skills**:

- ○ As AI automates routine tasks, there is an increasing demand for cognitive and emotional skills. A study by Bessen (2019) highlights this shift, noting a growing need for social, emotional, and higher cognitive skills in the workplace.

2. **Demand for Technological Proficiency**:
- ○ The integration of AI necessitates a workforce skilled in technology. Brynjolfsson and McAfee (2014) argue that workers must be proficient in AI-related technologies to remain competitive in the job market.

3. **Re-skilling and Up-skilling of Workers**:
- ○ The changing skill landscape requires significant re-skilling and up-skilling efforts. The World Economic Forum (2018) emphasizes the urgency of updating the skills of the current workforce to meet the demands of AI-driven economies.

4. **Adaptability and Lifelong Learning**:
- ○ The rapid evolution of AI technologies necessitates a focus on adaptability and lifelong learning. A report by McKinsey Global Institute (2017) underscores the importance of continuous learning to adapt to changing job requirements.

5. **Impact on Higher Education**:
- ○ Higher education institutions must evolve to meet these new skill demands. A study by Autor et al. (2020) discusses the need for universities to adapt their curricula to prepare students for an AI-driven world.

6. **Increase in Interdisciplinary Skills**:
- ○ The rise of AI has led to an increased demand for interdisciplinary skills, blending technology with other fields. This is supported by the research of Agrawal, Gans, and Goldfarb (2018), who highlight the importance of interdisciplinary knowledge in an AI-driven economy.

7. **Reduction in Demand for Low-Skill Jobs**:
   - AI and automation reduce the demand for low-skill, routine jobs. Acemoglu and Restrepo (2020) examine the displacement effects on low-skill workers, suggesting a need for policy interventions.
8. **New Job Creation**:
   - While AI displaces certain jobs, it also creates new ones. A report by the OECD (2019) explores how AI leads to the emergence of new occupations and skill requirements.
9. **Role of Private and Public Sector in Training**:
   - Both the private and public sectors play a crucial role in facilitating training and education for these new skill demands. Kaplan and Haenlein (2019) discuss the responsibilities of both sectors in supporting workforce transitions.

VI. Impact on Small and Medium Enterprises (SMEs):

AI offers SMEs opportunities to compete with larger enterprises by improving efficiency and access to markets. However, the cost of implementing AI can be a barrier for some SMEs (European Commission, 2018).

The integration of AI into the workforce presents unique challenges and opportunities for Small and Medium Enterprises (SMEs). Understanding these implications is crucial for both policy makers and business leaders.

1. **Increased Efficiency and Productivity**:
   - AI can significantly boost the efficiency and productivity of SMEs. A study by Kumar et al. (2019) suggests that AI integration can automate routine tasks, allowing SMEs to focus on strategic activities.
2. **Cost Reduction**:

- AI technologies can help SMEs reduce operational costs. Bughin et al. (2017) highlight how AI can streamline processes and reduce labor costs, making SMEs more competitive.

3. **Enhanced Customer Experience**:

- AI can improve customer service and experience. According to Gartner (2018), AI-driven customer service solutions can enhance customer engagement for SMEs.

4. **Barrier to Entry and Adoption**:

- The high cost and complexity of AI technologies can be a significant barrier for SMEs. A report by McKinsey Global Institute (2018) discusses the challenges SMEs face in adopting AI, including limited resources and expertise.

5. **Need for Skilled Workforce**:

- SMEs require a workforce skilled in AI and data analytics. A study by Deloitte (2019) emphasizes the importance of up-skilling employees to leverage AI technologies effectively.

6. **Market Competition and Innovation**:

- AI enables SMEs to compete with larger enterprises. The European Commission (2020) notes that AI can level the playing field, allowing SMEs to innovate and compete globally.

7. **Access to Data and Privacy Concerns**:

- SMEs must navigate data access and privacy issues. Research by the OECD (2021) highlights the challenges SMEs face in accessing quality data and complying with privacy regulations.

8. **Government Support and Policy**:

- Government policies play a crucial role in supporting SMEs in AI adoption. A study by the World Bank (2019) suggests that government initiatives can facilitate access to AI technologies for SMEs.

9. **Collaboration and Partnerships**:
   - Partnerships with technology providers and educational institutions can help SMEs adopt AI. The European Business Review (2020) discusses the benefits of collaborative efforts for AI integration in SMEs.

VII. Government Policy and Regulation:

Governments play a crucial role in shaping the impact of AI on the economy. Policies around education, retraining programs, and ethical guidelines for AI are essential (West, 2018).

Government policy and regulation play a pivotal role in shaping the economic implications of an AI workforce. This aspect involves balancing innovation with ethical considerations, workforce protection, and economic growth.

1. **Regulatory Frameworks for AI**:
   - Governments are tasked with creating regulatory frameworks that foster AI innovation while ensuring ethical use. The European Union's General Data Protection Regulation (GDPR) is a leading example, which, as discussed by Voigt, P., & Von dem Bussche, A. (2017), provides guidelines on AI-related data privacy.
2. **Workforce Protection and Transition Policies**:
   - Policies are needed to protect workers from job displacement due to AI. Autor, D., & Dorn, D. (2013) argue that such policies should include retraining programs and unemployment benefits to mitigate the impact of automation.
3. **Fostering AI Innovation and Competitiveness**:
   - Government investment in AI research and development is crucial for maintaining global competitiveness. A report by the National Science and Technology Council (2016)

highlights the importance of government funding in AI R&D.

4. **Ethical Guidelines and AI Governance:**

   ○ Establishing ethical guidelines for AI use is essential. Jobin, A., Ienca, M., & Vayena, E. (2019) discuss the need for governance frameworks that ensure AI is developed and used responsibly.

5. **Taxation and Revenue Policies:**

   ○ Governments must consider how AI impacts taxation and revenue. Brynjolfsson, E., & McAfee, A. (2014) discuss the challenges of taxing increasingly automated businesses and the potential for new forms of revenue generation.

6. **Education and Skill Development:**

   ○ Policies to promote education and skill development are vital in preparing the workforce for an AI-driven economy. A study by the World Economic Forum (2018) emphasizes the need for lifelong learning initiatives.

7. **Public-Private Partnerships:**

   ○ Collaborations between governments and private sector entities can accelerate AI advancement. A report by Accenture (2017) illustrates how public-private partnerships can drive innovation in AI.

8. **International Cooperation on AI Policy:**

   ○ International cooperation is critical for addressing global challenges posed by AI. As discussed by Taddeo, M., & Floridi, L. (2018), a global approach is necessary for effective AI governance.

VIII. Global Economic Shifts:

AI is influencing global economic dynamics, with countries investing in AI seeing potential shifts in economic power. This technological advancement could redefine global competitiveness (Lee, 2018).

The integration of Artificial Intelligence (AI) into the global workforce is causing significant shifts in the world's economic landscape. These shifts are characterized by changes in trade patterns, the emergence of new markets, and a redefinition of global economic power dynamics.

1. **Trade Patterns:**

- AI technologies are altering international trade by enabling more efficient production and logistics, thus impacting the comparative advantages of nations. For example, countries with advanced AI capabilities may develop new export sectors or improve existing ones. Brynjolfsson and McAfee (2014) discuss how technological advancements, including AI, influence economic productivity and trade, leading to a reshaping of global trade networks.

1. **Emergence of New Markets:**

- AI is creating new markets, particularly in areas like AI-driven analytics, robotics, and autonomous systems. According to a report by PwC (2017), AI is expected to contribute up to $15.7 trillion to the global economy by 2030, signaling the emergence of these new markets. This growth is attributed to AI-driven innovations and their applications across various sectors.

1. **Redefining Economic Power:**

- The nations that lead in AI technology are poised to gain significant economic power. For instance, the United States and China are currently leading the AI race, impacting their geopolitical and economic standings (Lee, 2018). This shift in power dynamics is partly due to the control over AI technologies and the ability to set standards in AI development and ethics.

1. **Impact on Developing Countries:**

- The rise of AI can both present opportunities and challenges for developing countries. While it offers potential for leapfrogging technological gaps, it also poses a risk of widening the digital divide. A study by the World Bank (2016) highlights the potential of AI to transform economies but also warns about the need for policies to ensure equitable benefits.

1. **Global Labor Markets:**

- The global labor market is undergoing transformations with AI automating certain jobs and creating new ones. Frey and Osborne (2017) predict that AI will lead to significant job displacement but also job creation in new areas, influencing labor migration and workforce development globally.

1. **Economic Policy and Cooperation:**

- Global economic policy and cooperation are becoming increasingly important in managing the impacts of AI. International forums like the G20 are beginning to focus on the implications of AI on trade, employment, and economic policies (G20 Summit, 2019). This highlights the need for coordinated efforts to harness the benefits of AI while mitigating its risks.

The potential for increased productivity and growth:

The integration of Artificial Intelligence (AI) into the workforce has the potential to significantly enhance productivity and drive economic growth. This potential is rooted in AI's ability to optimize operations, innovate processes, and create new business models.

1. **Enhanced Productivity:**

The beginning of AI in the workforce holds tremendous potential for enhanced productivity, a factor that is pivotal for economic growth. This enhancement is mainly attributed to AI's ability to automate routine tasks, improve efficiency, and support complex decision-making processes.

1. **Automation of Routine Tasks:**

- AI excels in handling repetitive and routine tasks, thus freeing human workers to focus on more complex and creative work. According to a report by Bughin, Hazan, and Ramaswamy (2018), AI can automate almost 45% of repetitive work, allowing employees to concentrate on tasks that add more value. This shift not only enhances productivity but also improves job satisfaction among employees.

1. **Efficiency in Operations:**

- AI technologies are known for their efficiency, especially in processing large volumes of data and performing tasks with precision. A study by Chui, Manyika, and Miremadi (2016) showed that AI could improve business process efficiency by 50-70%, thereby increasing overall organizational productivity.

1. **Improved Decision Making:**

- AI's capability in data analysis and predictive analytics enables better decision-making. According to Davenport and Ronanki (2018), AI tools can analyze vast datasets to provide insights that human analysis might miss, leading to more informed and strategic decisions.

1. Enhancing Worker Skills:

- AI not only replaces certain tasks but also enhances human capabilities. Kaplan and Haenlein (2019) argue that AI tools can augment human skills, leading to an overall increase in workforce productivity.

1. Challenges in Implementation:

- While the potential benefits are significant, implementing AI can be challenging. Issues such as the initial cost, need for technical expertise, and resistance to change are common obstacles. A report by the McKinsey Global Institute (2017) suggests that successful implementation of AI requires strategic planning and investment in human capital.

II. Innovation and New Business Models:

The integration of AI into the workforce is not just about enhancing productivity; it's a catalyst for innovation and the development of new business models. This transformative technology is reshaping industries by enabling novel approaches to solving problems and creating value.

1. Fostering Innovation:

- AI's ability to analyze and learn from data at unprecedented scales fosters a culture of innovation. For instance, Bessen (2019) noted that AI applications in research and development can significantly shorten the innovation cycle, leading to quicker advancement and deployment of new technologies.

1. Development of New Business Models:

- AI enables businesses to explore new models that were previously unfeasible. Huang, Rust, and Maksimovic (2019) highlight how AI-driven personalization has given rise to subscription-based and on-demand services, changing how companies interact with their customers.

1. **Enhancing Product Development:**

- AI's predictive analytics and machine learning capabilities allow companies to anticipate market trends and customer needs more accurately. According to Agrawal, Gans, and Goldfarb (2018), this leads to more efficient and targeted product development, reducing waste and improving market fit.

1. **Impact on Service Industries:**

- AI is revolutionizing service industries by automating customer interactions and personalizing services. A study by Mehta, Dahl, and Wattenhofer (2018) shows how AI-powered chatbots and recommendation systems enhance customer experience and engagement.

1. **Challenges and Considerations:**

- While AI presents vast opportunities, it also comes with challenges such as ethical considerations, the need for robust data, and the risk of obsolescence in traditional business models. Kaplan and Haenlein (2019) caution that businesses must navigate these challenges carefully to fully capitalize on AI's potential.

III. Economic Growth:

The integration of Artificial Intelligence (AI) into the workforce is a significant driver of economic growth. This growth is attributed to

AI's ability to enhance productivity, foster innovation, and create new market opportunities.

1. **Boosting Productivity:**

- AI technologies are known for their efficiency and speed in processing vast amounts of data, leading to increased productivity. A study by Bughin, Hazan, Ramaswamy, Chui, Allas, Dahlström, Henke, and Trench (2018) found that AI can potentially deliver additional global economic activity of around $13 trillion by 2030, which is about 1.2 percent additional GDP growth per year. This growth is largely driven by the automation of routine tasks and the augmentation of human capabilities.

1. **Catalyzing Innovation:**

- AI is a key driver of innovation, leading to the development of new products and services. According to Agrawal, Gans, and Goldfarb (2019), AI can transform existing business models and processes, creating new opportunities for economic growth.

1. **Enhancing Global Competitiveness:**

- The adoption of AI in industries can significantly enhance a nation's competitiveness on the global stage. A report by Manyika, Chui, Bughin, Woetzel, Dobbs, Roxburgh, and Byers (2017) emphasizes that countries leading in AI innovation and application are likely to gain a competitive edge in the global economy.

1. **Impact on Employment and Wages:**

- While AI drives economic growth, its impact on employment is multifaceted. Bessen (2019) argues that while AI can displace some jobs, it also creates new ones and increases the demand for AI-related skills, potentially leading to higher wages in certain sectors.

1. **Long-Term Economic Effects:**

- The long-term economic implications of AI are profound. As noted by Lee (2018), the economies that successfully integrate AI will see significant growth in productivity and GDP. However, there is also a need to address the challenges posed by AI, including its impact on the labor market and income inequality.

IV. Reducing Operational Costs:

The adoption of Artificial Intelligence (AI) in various sectors is significantly reducing operational costs, contributing to increased productivity and economic growth. AI's ability to streamline processes, optimize resource usage, and improve decision-making processes plays a crucial role in cost reduction.

1. **Streamlining Processes:**

- AI technologies are adept at automating routine tasks, which reduces the time and resources spent on such activities. A study by Fountaine, McCarthy, and Saleh (2019) highlights that companies employing AI for process automation report a reduction in operational costs. This efficiency not only cuts costs but also reallocates human labor to more strategic tasks, thereby enhancing productivity.

1. **Enhancing Resource Optimization:**

- AI systems are capable of analyzing large datasets to identify patterns and make predictions, leading to more efficient resource management. According to Davenport and Ronanki (2018), AI-driven analytics can significantly reduce waste and optimize resource allocation, resulting in cost savings for businesses.

1. **Improving Decision-Making:**

- AI aids in making more informed decisions by providing insights based on data analysis. Bughin, Seong, Manyika, Chui, and Joshi (2018) note that AI's predictive capabilities enable businesses to make strategic decisions that reduce costs and improve efficiency.

1. **Maintenance and Operations:**

- AI technologies, particularly in the field of predictive maintenance, can foresee potential equipment failures, allowing for timely maintenance and reducing downtime costs. A report by the McKinsey Global Institute (2017) states that predictive maintenance can reduce machine downtime by up to 50% and lower maintenance costs by 20-25%.

1. **Energy Savings:**

- AI's ability to optimize energy usage in industrial processes and buildings significantly cuts energy costs. A study by Wei, Chan, Wang, and Cai (2019) found that AI-driven energy optimization can lead to substantial reductions in energy consumption, thereby reducing operational expenses.

1. **Customer Service Enhancement:**

- AI-powered chatbots and virtual assistants reduce the cost of customer service operations while improving service quality. As

Kapoor, Lee, and Nair (2018) explain, these AI tools handle routine customer queries efficiently, reducing the need for a large customer service workforce.

V. Job Creation:

The integration of AI into the workforce, while often associated with job displacement, also holds significant potential for job creation. This facet is critical in understanding the comprehensive economic implications of AI.

1. Emergence of New Job Categories:

- AI and automation lead to the creation of new job roles that did not exist before. Bessen (2019) notes that technology historically has created more jobs than it destroys, with AI expected to follow this trend. These new roles often require skills in AI management, development, and oversight.

1. Augmenting Existing Jobs:

- Rather than replacing human workers, AI often augments their roles, leading to job transformation rather than elimination. A report by World Economic Forum (2018) suggests that AI will create 58 million new jobs by 2022, as it changes the nature of existing jobs and creates demand for new skills.

1. Increasing Demand in Tech-Related Fields:

- The rise of AI significantly boosts the demand for professionals in data science, machine learning, and AI ethics. According to a study by Frank, Roehrig, and Pring (2019), there is a burgeoning demand for roles such as AI specialists and data analysts.

1. **Expanding the Service Sector:**

- AI's capabilities in personalization and efficiency can lead to an expansion of the service sector. Kaplan and Haenlein (2019) argue that personalized AI-driven services create new market opportunities, thus generating new employment in these areas.

1. **Enhancing Human Capabilities:**

- AI tools enable workers to perform tasks more efficiently, potentially leading to job growth in sectors where human judgment and creativity are irreplaceable. A study by Agrawal, Gans, and Goldfarb (2018) indicates that AI can enhance human capabilities, leading to growth in jobs that require these unique human skills.

1. **Indirect Job Creation:**

- The deployment of AI in various industries can stimulate economic growth, leading to the creation of jobs indirectly related to AI. According to a report by PwC (2017), AI is expected to contribute $15.7 trillion to the global economy by 2030, potentially creating numerous indirect employment opportunities.

VI. Challenges and Limitations

While the potential for increased productivity and economic growth through AI is significant, it's essential to recognize the challenges and limitations that accompany this technological advancement.

1. **Skill Gap and Workforce Displacement:**

- One of the foremost challenges is the displacement of workers whose skills are rendered obsolete by AI. Autor, Levy, and

Murnane (2003) highlight the growing need for skills that complement technology, suggesting a gap between current workforce capabilities and future demands. This necessitates significant investment in retraining and upskilling.

1. **Ethical and Societal Concerns:**

- The implementation of AI raises ethical questions, including privacy, bias, and accountability. Bostrom (2014) discusses the ethical challenges posed by AI, emphasizing the need for robust frameworks to guide AI development and implementation.

1. **Economic Inequality:**

- AI could exacerbate economic inequality. Brynjolfsson and McAfee (2014) warn that AI-driven growth may disproportionately benefit those with the skills and capital to leverage it, widening the income gap.

1. **Dependence on Data and Privacy Issues:**

- AI systems require large amounts of data, raising concerns about privacy and data security. Mayer-Schönberger and Cukier (2013) note the potential for misuse of data, emphasizing the importance of establishing stringent data protection regulations.

1. **Limitations in AI's Capabilities:**

- While AI can process and analyze data at an unprecedented scale, it is not without limitations. Marcus (2018) points out that current AI technologies, particularly those based on machine learning, have constraints in understanding context and generalizing beyond their training data.

1. **Infrastructure and Investment Requirements:**

- The implementation of AI requires significant infrastructure and investment. A report by McKinsey Global Institute (2017) states that countries and businesses must invest heavily in AI technologies and infrastructure to reap the economic benefits, which might not be feasible for all.

1. **Global Competition and Technological Dominance:**

- The race for AI dominance can lead to global disparities. Lee (2018) discusses the geopolitical implications of AI, with the potential for a few AI superpowers to dominate the technology, leaving others behind.

Addressing wealth inequality in the AI epoch:

The rise of AI in the workforce presents unique challenges and opportunities concerning wealth inequality. Addressing this inequality in the AI era requires a multifaceted approach, involving both policy and private sector initiatives.

1. **Progressive Taxation and Redistribution:**

Progressive Taxation and Redistribution is a key strategy for addressing wealth inequality in the age of AI. This approach involves adjusting the tax system to ensure that those benefiting most from AI-driven economic growth contribute a fairer share to the public purse. The redistributed funds are then used to support social programs, infrastructure, and initiatives that benefit the wider society, especially those impacted by AI-related job displacement.

1. **Rationale for Progressive Taxation in the AI Era:**

- Piketty (2014) argues that as AI and automation increase productivity and profits, the resulting wealth is often concentrated in the hands of a few. Progressive taxation aims to address this by imposing higher tax rates on higher income brackets, particularly targeting income generated through AI and technology-driven ventures.

1. **Redistribution for Social Equity:**

- The revenues generated from progressive taxation can be channeled into social programs such as healthcare, education, and unemployment benefits. This helps in cushioning the workforce against the disruptive effects of AI, as highlighted by Saez and Zucman (2019), who propose using tax revenues to fund universal healthcare and education programs.

1. **Addressing Job Displacement:**

- Brynjolfsson and McAfee (2014) suggest that the increased efficiency brought about by AI will lead to significant job displacement. Redistribution can fund retraining and reskilling programs, helping workers transition to new roles in an AI-driven economy.

1. **Ensuring Fair Contribution by Tech Giants:**

- Stiglitz (2019) emphasizes the importance of ensuring that companies profiting from AI and data pay their fair share of taxes. This includes closing loopholes and ensuring global cooperation to prevent tax avoidance.

1. **Economic Stability and Growth:**

- Redistribution through progressive taxation can lead to more equitable economic growth. As Atkinson (2015) points out, reducing inequality through such means can stimulate economic stability and growth, as a more equitable distribution of wealth can lead to increased consumer spending and investment in human capital.

1. **Political and Social Considerations:**

- The implementation of progressive taxation and redistribution in the context of AI raises political and social considerations. It requires careful policy design to balance the interests of various stakeholders, including businesses, workers, and consumers.

II. Universal Basic Income (UBI):

Universal Basic Income (UBI) has emerged as a proposed solution to the challenges of wealth inequality exacerbated by the AI revolution. UBI involves providing all citizens with a regular, unconditional sum of money, regardless of employment status or income. The rationale is to ensure a basic standard of living, thereby reducing poverty and inequality.

1. **UBI as a Response to AI-induced Job Displacement:**

- Bregman (2017) argues that UBI is a necessary response to the job displacement caused by AI and automation. As machines replace human labor, UBI can provide financial security to those who lose their jobs or find their skills obsolete.

1. **Economic Security and Consumer Spending:**

- Standing (2017) suggests that UBI can contribute to economic stability by ensuring that people have enough money to meet their basic needs, which in turn can stimulate consumer spending and economic growth.

1. **Promoting Innovation and Entrepreneurship:**

- According to Yang (2018), UBI can encourage entrepreneurship by reducing the financial risks associated with starting new businesses. It provides a safety net that allows individuals to innovate and take entrepreneurial risks without the fear of financial ruin.

1. **Addressing Inequality:**

- Van Parijs and Vanderborght (2017) highlight that UBI can be a tool for reducing income and wealth inequality. By providing a basic income, it ensures a minimum standard of living for everyone, narrowing the gap between the rich and the poor.

1. **Challenges in Implementation:**

- Critics, such as Lowrey (2018), point out the challenges in implementing UBI, including the high costs and potential disincentives to work. Determining an appropriate level of UBI that is both sustainable and effective in reducing inequality is a key challenge.

1. **Funding UBI in the AI Era:**

- Hughes (2018) discusses various models for funding UBI, including taxation of AI-driven enterprises. Taxing the profits from AI and automation could provide a revenue stream to fund UBI, linking the wealth generated by AI directly to the solution for inequality.

1. Social and Psychological Benefits:

- UBI can also have positive social and psychological effects. As noted by Susskind (2020), beyond economic benefits, UBI can lead to improvements in mental health, community cohesion, and overall quality of life.

III. Education and Reskilling Initiatives:

Education and reskilling initiatives are crucial in addressing the wealth inequality intensified by the advent of AI in the workforce. These programs aim to equip individuals with the skills necessary to thrive in an increasingly automated job market.

1. Bridging the Skill Gap:

- Autor (2015) emphasizes the importance of education and vocational training in bridging the skill gap created by AI and automation. Reskilling workers for high-demand areas can mitigate the impact of job displacement and ensure a more equitable distribution of the new job opportunities created by AI.

1. Lifelong Learning as a Norm:

- Schwab (2016) advocates for the concept of lifelong learning, where continuous education becomes a norm throughout an individual's career. This approach helps workers stay relevant and adaptable in a rapidly changing job market.

1. Public-Private Partnerships in Reskilling:

- Bessen (2019) highlights the role of public-private partnerships in successful reskilling initiatives. Collaboration between governments, educational institutions, and corporations can lead to more effective and targeted training programs.

1. **Economic Benefits of Reskilling:**

- A study by McKinsey Global Institute (2018) found that re-skilling workers can not only reduce inequality but also add significant value to the economy by improving the skill base of the workforce.

1. **Accessibility and Inclusion in Education Programs:**

- Moretti (2012) stresses the need for these educational initiatives to be accessible and inclusive, ensuring that marginalized groups are not left behind in the shift towards a more AI-driven economy.

1. **Challenges in Reskilling:**

- Autor and Dorn (2013) acknowledge the challenges in reskilling, including the time and financial resources required, and the need for these initiatives to be aligned with the evolving demands of the labor market.

1. **Policy Recommendations for Supporting Reskilling:**

- The World Economic Forum (2020) provides policy recommendations for supporting reskilling efforts, including government subsidies for education, tax incentives for companies investing in employee training, and the development of online platforms for digital learning.

IV. Regulation of AI and Data Monopolies:

Regulating AI and data monopolies is essential in addressing wealth inequality in the era of artificial intelligence. This approach focuses on ensuring fair competition, protecting consumer rights, and preventing the concentration of wealth and power in the hands of a few dominant players.

1. **Promoting Competition and Innovation:**

- Stiglitz (2019) argues for the need to regulate monopolies in the AI sector to promote competition and innovation. Antitrust laws and regulations can prevent a few firms from dominating the market, which can stifle innovation and exacerbate wealth inequality.

1. **Data Privacy and Consumer Rights:**

- The European Union's General Data Protection Regulation (GDPR), implemented in 2018, provides a framework for protecting consumer data and privacy (European Commission, 2018). This regulation ensures that individuals have control over their personal data, which is essential in an economy increasingly driven by big data and AI.

1. **Economic Impacts of Data Monopolies:**

- A report by the OECD (2019) highlights the economic impacts of data monopolies, noting that they can lead to market imbalances and inequalities. The report recommends regulatory interventions to ensure a level playing field and to protect smaller players in the market.

1. **Addressing Wealth Concentration:**

- Piketty (2014) discusses the broader issue of wealth concentration and suggests that regulation of large corporations, including AI and data companies, is necessary to prevent the exacerbation of economic inequality.

1. **Global Cooperation in AI Regulation:**

- Lee (2018) emphasizes the importance of global cooperation in regulating AI. As AI technologies cross borders, international regulatory frameworks are needed to manage their impact on global wealth distribution.

1. **Regulatory Challenges and Balance:**

- Furman and Seamans (2019) acknowledge the challenges in regulating AI and data monopolies. They suggest a balanced approach that avoids stifling innovation while ensuring fair competition and consumer protection.

1. **Impact on Small and Medium Enterprises (SMEs):**

- A study by the World Bank (2020) points out that regulation of AI and data monopolies can also benefit SMEs by creating a more competitive environment that allows them to thrive alongside larger corporations.

V. Inclusive AI Development:

Inclusive AI development is pivotal in addressing wealth inequality in the AI epoch. This approach involves creating AI technologies that are accessible and beneficial to all segments of society, thereby reducing the digital divide and ensuring equitable distribution of AI's economic benefits.

1. **Broadening Access to AI Technologies:**

- The World Bank (2019) emphasizes the importance of broadening access to AI technologies to ensure inclusive development. This involves not only making AI tools available but also ensuring that diverse populations have the skills to use and benefit from these technologies.

1. **Participatory Design in AI:**

- A study by Shneiderman (2016) highlights the role of participatory design in AI, where end-users are involved in the development process. This ensures that AI systems cater to the needs of a wider range of people, thereby reducing the risk of exacerbating existing inequalities.

1. **Impact on Marginalized Communities:**

- Research by Eubanks (2018) on the use of AI in public services shows how AI can either perpetuate or reduce inequalities, depending on its implementation. Inclusive AI development requires consideration of the impacts on marginalized communities and the proactive addressing of potential biases.

1. **AI for Social Good:**

- The United Nations (2018) report on "AI for Social Good" discusses how AI can be leveraged to address global challenges like poverty, hunger, and inequality. It suggests that AI development should align with the Sustainable Development Goals to ensure it contributes to reducing wealth disparities.

1. **Diversity in AI Workforce:**

- A study by West et al. (2019) stresses the importance of diversity in the AI workforce. It argues that a diverse group of AI researchers and developers can create more inclusive AI systems that are sensitive to the needs of different societal groups.

1. **Ethical AI Development:**

- The European Commission's High-Level Expert Group on Artificial Intelligence (2019) proposes ethical guidelines for AI development. These guidelines include ensuring that AI systems are fair, non-discriminatory, and serve the interests of all, especially the most vulnerable.

1. **Public-Private Partnerships in AI:**

- Lee (2020) suggests that public-private partnerships can play a key role in inclusive AI development. By collaborating, governments and technology firms can ensure that AI advancements are accessible and beneficial to a broader population.

VI. Public-Private Partnerships:

1. **Rationale for PPPs**: Public-private partnerships are emerging as a vital tool in addressing wealth inequality in the era of AI-driven economic change. These partnerships can leverage the strengths of both sectors to address complex social issues, including wealth disparity exacerbated by AI and automation (Schwab & Davis, 2018).
2. **Case Studies and Evidence**:
    - **Innovative Training Programs**: For instance, Microsoft's collaboration with community colleges in the U.S. to offer AI-related training is a notable example. This initiative aims to provide accessible education and reskilling

opportunities to a broad demographic, helping bridge the inequality gap (Microsoft, 2020).

- **Research and Development Investments**: Joint investments in AI research can lead to more equitable AI solutions. An example is the partnership between the European Commission and private entities in funding AI research that focuses on ethical and inclusive technology (European Commission, 2021).

3. **Benefits of PPPs**:

- **Enhanced Resource Utilization**: By combining public funding and oversight with private sector innovation and efficiency, PPPs can more effectively address inequality issues (Kolk, Van Dolen, & Vock, 2020).

- **Increased Access to Technology**: Through these partnerships, underserved communities can gain better access to AI technologies and training, thereby reducing the digital divide (ITU, 2019).

4. **Challenges and Considerations**:

- **Balancing Interests**: Ensuring that both public and private entities benefit equitably from PPPs is crucial. This balance is necessary to maintain long-term, sustainable partnerships (Hodge & Greve, 2019).

- **Regulatory Frameworks**: Governments need to establish clear regulatory frameworks to guide PPPs in AI, ensuring that these initiatives align with broader societal goals, such as reducing wealth inequality (OECD, 2021).

5. **Future Outlook**:

- **Expanding the Scope of PPPs**: As AI continues to evolve, there is a growing need for PPPs to address not just training and education but also ethical AI development and the creation of inclusive AI policies (WEF, 2020).

6. **Conclusion**:

○ Public-private partnerships offer a promising avenue for addressing wealth inequality in the AI epoch. By leveraging the strengths of both sectors, these partnerships can foster inclusive growth and ensure that the benefits of AI are widely distributed (Schwartz et al., 2019).

VII. Global Cooperation:

1. **Importance of Global Cooperation**: In the context of AI-driven economic transformations, global cooperation emerges as a critical strategy for addressing wealth inequality. This approach involves collaboration among nations to establish equitable AI policies and practices (Lee, 2018).

2. **Examples and Evidence**:

   ○ **International AI Agreements**: Agreements like the OECD's AI Principles, which many countries have endorsed, aim to promote AI's responsible stewardship to benefit society and reduce inequality (OECD, 2019).

   ○ **Cross-Border Data Sharing**: Initiatives like the EU's General Data Protection Regulation (GDPR) demonstrate how international standards can shape the ethical use of AI, potentially reducing wealth inequality caused by data monopolies (European Commission, 2018).

3. **Benefits of Global Cooperation**:

   ○ **Standardized AI Ethics and Guidelines**: By collaborating, countries can develop and adopt universal ethical standards for AI, which can prevent the exploitation of vulnerable populations (Jobin, Ienca, & Vayena, 2019).

   ○ **Sharing of Best Practices**: Countries can learn from each other's experiences in implementing AI in a way that mitigates inequality, such as through inclusive education systems and workforce development (UNESCO, 2021).

4. **Challenges and Considerations**:

- ○ **Diverse Economic and Political Contexts**: Achieving consensus among nations with varying economic and political landscapes is challenging, potentially hindering the effectiveness of global cooperation (Bostrom & Dafoe, 2021).
- ○ **Technology Transfer and Intellectual Property Issues**: Balancing the need for open sharing of AI technology with intellectual property rights is a complex aspect of international cooperation (WIPO, 2020).

5. **Future Outlook**:

- ○ **Expanding International AI Frameworks**: Future efforts may involve expanding existing frameworks or creating new international bodies dedicated to AI governance to further tackle issues of inequality (Taddeo & Floridi, 2018).

6. **Conclusion**:

- ○ Global cooperation presents a formidable approach to addressing the wealth inequality challenges posed by the rise of AI. Through collective effort and shared principles, nations can work towards a more equitable AI future (Russell, 2019).

Economic policies and strategies for a balanced AI-driven economy:

1. **Introduction to Economic Policies in the AI Era**:

- ○ With the advent of AI in the workforce, governments and policymakers need to adopt novel economic strategies to ensure a balanced and equitable economy (Acemoglu & Restrepo, 2020).

2. **Key Economic Policies and Strategies**:

- ○ **Investment in Education and Training**: Governments should prioritize education systems that focus on skills complementary to AI, including critical thinking and creativity (Brynjolfsson & McAfee, 2014).

- ○ **Taxation Policies**: Implementing taxation strategies on AI-driven businesses can help redistribute wealth and fund public services. This includes revising corporate tax structures and considering taxes on AI robots (Frey, 2019).
- ○ **Universal Basic Income (UBI)**: UBI has been proposed as a solution to offset job losses due to AI automation, providing a safety net for all citizens (Piketty, 2020).

3. **Strategies for SMEs and Startups**:
   - ○ **Support for SMEs and Startups**: Encouraging the adoption of AI in small and medium enterprises through subsidies or tax incentives can boost innovation and competitiveness (Autor, 2015).

4. **Fostering Innovation and Competition**:
   - ○ **Regulating AI Monopolies**: Implementing antitrust laws to prevent AI monopolies and encourage competition is crucial for a balanced economy (Zuboff, 2019).
   - ○ **Public-Private Partnerships**: Collaboration between the public sector and private enterprises in AI development can spur innovation while ensuring ethical AI deployment (Lee, 2018).

5. **Global Cooperation and Trade Policies**:
   - ○ **International Collaboration**: Establishing international agreements on AI standards and trade can promote ethical AI practices globally (Taddeo & Floridi, 2018).
   - ○ **Trade Policies**: Adjusting trade policies to consider the impact of AI on global supply chains and employment is necessary (Baldwin, 2016).

6. **Addressing Inequality and Social Welfare**:

- ○ **Social Welfare Programs**: Enhancing social welfare programs to support workers displaced by AI is vital for mitigating inequality (Atkinson, 2015).

7. **Conclusion**:

- ○ To harness the benefits of an AI-driven economy while mitigating its risks, a multifaceted approach involving education, taxation, regulation, and international cooperation is essential (Acemoglu & Restrepo, 2020).

13

Chapter 8: Global Perspectives on AI Employment

1. **Introduction:**
   - The integration of AI in the workforce presents diverse perspectives and challenges on a global scale, with varying impacts in different regions (Frey, 2019).

2. **North America's Embrace of AI:**
   - In the United States and Canada, there is a strong trend towards integrating AI into various sectors such as healthcare, finance, and technology, driven by significant investments and a strong technological infrastructure (Brynjolfsson & McAfee, 2014).

3. **Europe's Focus on Ethical AI:**
   - European countries, led by the European Union, are emphasizing the ethical dimensions of AI employment, focusing on regulations that protect workers and ensure fair AI practices (Floridi & Cowls, 2019).

4. **Asia's Rapid AI Adoption:**

- Asian countries, notably China and South Korea, are rapidly adopting AI in manufacturing and digital services, investing heavily in AI education and research (Lee, 2018).

5. **AI's Impact in Africa:**

- African nations face unique challenges due to limited technological infrastructure, but there is potential for AI to address issues such as healthcare delivery and agricultural efficiency (Taddeo & Floridi, 2018).

6. **Latin America's Developing AI Landscape:**

- In Latin America, AI is still in the developing stages with growing interest in adopting AI for improving public services and enhancing economic competitiveness (Brynjolfsson & McAfee, 2014).

7. **Challenges in Middle East:**

- The Middle East faces the challenge of integrating AI while balancing traditional industries with new technology, with countries like the UAE leading in AI investment (Hajkowicz et al., 2019).

8. **Australia's AI Strategy:**

- Australia is focusing on leveraging AI for improving healthcare, environmental management, and mining, with emphasis on ethical AI deployment (Hajkowicz et al., 2019).

9. **Conclusion:**

- Global perspectives on AI employment are diverse, reflecting regional economic, cultural, and technological landscapes. Balancing economic benefits with ethical considerations is key to successful AI integration (Acemoglu & Restrepo, 2020).

A tour of AI workplace impacts around the world:

1. **Introduction:**

- The impact of AI on the global workforce is multifaceted, varying significantly across different regions and industries (Schwab, 2017).

2. **North America: The Tech-Driven Workforce Revolution**:
   - In the United States and Canada, AI is largely reshaping the tech industry, with Silicon Valley being a prime example. The region has seen a surge in demand for AI-related skills, significantly impacting job dynamics in the tech sector (Brynjolfsson & McAfee, 2016).

3. **Europe: Balancing AI and Human Labor**:
   - European countries are focusing on balancing AI advancements with human labor. Initiatives like Germany's Industrie 4.0 represent efforts to harmonize AI with traditional manufacturing, emphasizing worker retraining and education (Renda, 2019).

4. **Asia: The Automation Frontier**:
   - Asian countries, particularly Japan and South Korea, are embracing automation and AI in industries ranging from manufacturing to service. In Japan, AI and robotics are being integrated into sectors like elder care, addressing demographic challenges (Lee, 2018).

5. **Africa: AI for Developmental Challenges**:
   - In Africa, AI is being utilized to tackle developmental challenges. For instance, in agriculture, AI is employed for crop monitoring and predictive analysis, aiding small-scale farmers in countries like Kenya (Heeks, 2018).

6. **Latin America: Emerging AI Markets**:
   - Latin American countries are emerging markets for AI, with nations like Brazil and Mexico beginning to adopt AI in sectors such as banking and retail, fostering economic growth and modernization (Castro, McLaughlin, & Chivot, 2019).

7. **Middle East: AI in Oil and Beyond**:

- In the Middle East, countries like the UAE are leveraging AI not only in traditional sectors like oil but also in public services and healthcare, aiming to become regional AI hubs (Hajkowicz et al., 2019).

8. **Australia and Oceania: AI in Environmental Management**:

- Australia is utilizing AI in environmental management and conservation efforts, applying AI technologies for wildlife protection and natural resource management (Bradley & Lawrence, 2019).

9. **Conclusion**:

- This global tour of AI's impact on the workplace reveals diverse applications and challenges, demonstrating how different regions adapt AI technologies to their unique economic and social contexts (Schwab, 2017).

Countries leading in AI and their employment strategies:

1. **Introduction**:

- The race to lead in AI is not just technological but also deeply tied to employment strategies. Key players in AI, such as the United States, China, and European countries, have developed unique approaches to integrate AI into their workforce (Susskind & Susskind, 2015).

2. **United States: Innovation and Entrepreneurship**:

- The U.S. leads in AI primarily through innovation and entrepreneurship. The focus is on cultivating AI startups and tech giants, promoting job creation in cutting-edge AI research and development. This strategy emphasizes the creation of high-skilled jobs and encourages continuous learning and adaptation among workers (Brynjolfsson & McAfee, 2016).

3. **China: Government-Led AI Development**:

- China adopts a government-led approach to AI, aiming to become the world leader in AI by 2030. The Chinese strategy involves heavy investment in AI education and training, aiming to prepare a vast workforce for AI-driven industries. This approach is also characterized by significant state investment in AI companies and research initiatives (Lee, 2018).

4. **European Union: AI for Social Good**:

   - The European Union focuses on leveraging AI for social good, integrating ethical and sustainable practices into AI development. European employment strategies in AI involve heavy investment in education and vocational training, aiming to prepare workers for an AI-driven economy while ensuring social welfare and job security (Renda, 2019).

5. **Japan: Addressing Demographic Challenges**:

   - Japan uses AI to counter its aging population and shrinking workforce. Their strategy involves integrating AI and robotics in various sectors, including healthcare and manufacturing, to maintain productivity levels. Japan also invests in AI education and training programs, focusing on both younger generations and retraining older workers (Frey & Osborne, 2017).

6. **India: Focusing on AI Skills Training**:

   - India, with its vast pool of young talent, focuses on AI skills training. The government and private sector collaborate to provide AI education and training programs, aiming to make India a significant player in the global AI market, particularly in software and service sectors (Rajan, 2020).

7. **Canada: AI Research and Ethical AI**:

   - Canada is known for its contributions to AI research, particularly in deep learning. The Canadian AI strategy includes significant investments in higher education and research institutes. Canada also emphasizes ethical AI development, aiming to create a responsible and inclusive AI workforce (Crawford & Calo, 2016).

8. **Conclusion**:

   ○ Each leading country in AI adopts a unique employment strategy, reflecting its socio-economic priorities and challenges. These strategies provide insights into how nations can harness AI for economic growth while addressing workforce challenges (Schwab, 2017).

Lessons from diverse economies and cultures:

1. **Introduction**:

   ○ The integration of AI into the workforce varies significantly across different economies and cultures, offering valuable lessons on adaptation, policy, and societal impact (Bughin, Seong, Manyika, Chui, & Joshi, 2018).

2. **Germany: Skilled Workforce and Automation**:

   ○ Germany's focus on high-quality manufacturing and a highly skilled workforce offers lessons in integrating AI with existing industrial strengths. The country demonstrates how AI can augment rather than replace skilled labor, emphasizing the importance of continuous training and education (Autor, Levy, & Murnane, 2003).

3. **South Korea: Rapid Adoption and Government Support**:

   ○ South Korea's rapid adoption of AI, supported by substantial government investment, shows the effectiveness of a coordinated national strategy. This approach emphasizes not only technological development but also social safety nets and retraining programs for displaced workers (Choi, 2020).

4. **Nigeria: Leveraging AI in Emerging Economies**:

- Nigeria, representing emerging economies, demonstrates how AI can be leveraged to leapfrog development stages. The focus on AI in areas like agriculture and healthcare shows the potential for AI to drive growth and address local challenges in developing economies (Okeleke & Suleyman, 2019).

5. **Brazil: AI in Resource-Rich Economies**:

   - Brazil's approach to AI in a resource-rich economy underscores the potential for AI to enhance resource management and environmental sustainability. Brazil's use of AI in sectors like agriculture and mining indicates the role AI can play in optimizing the use of natural resources (da Silva, Neto, & Lima, 2021).

6. **Singapore: A Model of AI Governance**:

   - Singapore's model of AI governance, focusing on ethical and responsible AI development, offers lessons in regulatory frameworks. This approach highlights the importance of trust and transparency in AI deployment, particularly in sectors like finance and public services (Tham, 2018).

7. **India: Balancing AI and Large Workforce**:

   - India's challenge of integrating AI in a country with a large workforce demonstrates the balance between embracing technology and safeguarding employment. India's approach focuses on upskilling and reskilling, highlighting the role of education in transitioning to an AI-driven economy (Kapoor, 2019).

8. **Conclusion**:

   - The diverse approaches to AI employment across different economies and cultures offer a broad spectrum of lessons, from policy formulation to societal adaptation. These insights are crucial for countries navigating the AI transformation in their unique socio-economic contexts (Brynjolfsson & McAfee, 2016).

14

Appendices

Glossary of AI and employment-related terms:

1. **Artificial Intelligence (AI):**

   A field of computer science dedicated to the creation of systems capable of performing tasks that typically require human intelligence, such as visual perception, speech recognition, decision-making, and language translation.

2. **Automation:**

   The technology by which a process or procedure is performed with minimal human assistance, often used in the context of replacing manual labor with machine-operated tasks.

3. **Machine Learning (ML):**

   A subset of AI that involves the development of algorithms that can learn and make predictions or decisions based on data.

4. **Deep Learning:**

   An advanced type of machine learning involving neural networks with many layers, enabling the computer to identify patterns and make decisions with minimal human intervention.

5. **Robotics:**

   The branch of technology that deals with the design, construction, operation, and application of robots, often used in manufacturing, healthcare, and other industries.

6. **Natural Language Processing (NLP)**:

A field of AI that focuses on the interaction between computers and human language, enabling computers to understand, interpret, and respond to human language in a useful way.

7. **Algorithm**:

A set of rules or instructions given to an AI program to help it learn and make decisions.

8. **Data Mining**:

The process of discovering patterns and extracting valuable information from large sets of data, often used in AI to improve decision-making and predictions.

9. **Unemployment**:

The condition of being jobless, often used in the context of discussions about the impact of AI and automation on the job market.

10. **Reskilling**:

The process of learning new skills so one can do a different job, or of training people to do a different job, often discussed in the context of workforce adaptation to AI advancements.

11. **Gig Economy**:

A labor market characterized by the prevalence of short-term contracts or freelance work as opposed to permanent jobs, often influenced by digital platforms and AI technologies.

12. **Universal Basic Income (UBI)**:

A model for providing all citizens with a given sum of money, regardless of their income, resources, or employment status, as a way to address unemployment and inequality in the AI era.

13. **Ethical AI**:

The practice of designing, developing, and deploying AI systems in a way that respects ethical principles and values such as fairness, accountability, and transparency.

14. **AI Governance**:

The strategies, policies, and regulations that govern the development and use of AI technologies in society.

15. **Digital Divide**:

The gap between demographics and regions that have access to modern information and communications technology, and those that don't or have restricted access, particularly relevant in discussions of AI's impact on global workforce dynamics.

16. **Industry 4.0**:

The current trend of automation and data exchange in manufacturing technologies, including cyber-physical systems, the Internet of things, cloud computing, and cognitive computing, often synonymous with the fourth industrial revolution.

This glossary provides a foundational understanding of key terms related to AI and its impact on employment, offering a useful reference for discussions in this field.

Overview of key AI technologies influencing job markets:

1. **Machine Learning (ML)**:

ML algorithms analyze data to learn patterns and make decisions or predictions. In the job market, ML impacts roles in data analysis, software development, and research sectors by automating complex data-driven tasks.

2. **Natural Language Processing (NLP)**:

NLP technology enables machines to understand and interpret human language. Its influence is seen in customer service through chatbots, in legal and healthcare documentation, and in content creation industries.

3. **Robotics**:

Robotics technology automates manual tasks, significantly

impacting manufacturing, logistics, and even healthcare. Robots in factories can improve efficiency but may replace certain manual jobs.

4. **Computer Vision**:

This technology, which enables machines to interpret and process visual data, is transforming industries like retail, security, and transportation. For example, in retail, computer vision assists in inventory management and customer service.

5. **Predictive Analytics**:

Utilizing AI to predict trends and behavior patterns, predictive analytics is reshaping marketing, finance, and operations planning. It helps businesses anticipate market trends, influencing demand for data scientists and analysts.

6. **AI in Healthcare**:

AI applications in diagnostics, patient care, and personalized medicine are revolutionizing the healthcare sector. They demand new skills from medical professionals but also create efficiency in patient management and research.

7. **Autonomous Vehicles**:

Self-driving technology impacts the transportation sector, potentially reducing the need for drivers but increasing demand for engineers and technicians skilled in AI and robotics.

8. **AI in Finance**:

AI-driven algorithms for risk assessment, fraud detection, and customer service automation are reshaping banking and finance, impacting roles in customer service, risk management, and financial analysis.

9. **Speech Recognition**:

Used in devices like smart speakers and in customer service automation, speech recognition technology is influencing jobs in technology development, customer support, and data entry fields.

10. **Internet of Things (IoT):**

    IoT, combined with AI, is leading to smarter, interconnected devices, impacting jobs in software development, system architecture, and data analysis, particularly in industries like home automation and smart cities.

11. **AI in Education:**

    AI-driven personalized learning and administrative automation are changing the education sector. While they augment teaching methods, they also demand tech-savvy educators and administrators.

12. **AI in Agriculture:**

    AI applications in crop monitoring and predictive analysis are altering agricultural jobs, leading to a decrease in manual labor but an increase in demand for tech specialists in agriculture.

13. **Ethical AI & Governance:**

    As AI becomes more prevalent, the need for ethical guidelines and governance increases, influencing legal, policy-making, and compliance jobs.

14. **AI-Enabled Cybersecurity:**

    AI-driven security systems are crucial in protecting data, impacting jobs in cybersecurity, where there is a growing need for professionals skilled in AI-based security protocols.

This overview highlights how various AI technologies are influencing job markets across different sectors, creating a dynamic where some roles are diminished, others are transformed, and new ones are created.

Resources for further reading and exploration:

1. **Books:**

- *"AI Superpowers: China, Silicon Valley, and the New World Order" by Kai-Fu Lee*: Offers insights into the global AI race and its implications on the future job market.
- *"Humans Need Not Apply: A Guide to Wealth and Work in the Age of Artificial Intelligence" by Jerry Kaplan*: Explores the changes AI will bring to the economy and job market.
- *"The Industries of the Future" by Alec Ross*: Discusses the future of various industries in the context of AI and other advancing technologies.

2. **Research Papers and Journals**:

- *Journal of Artificial Intelligence Research*: A source for the latest research findings in AI.
- *"The Future of Employment: How Susceptible Are Jobs to Computerisation?" by Carl Benedikt Frey and Michael A. Osborne*: A seminal paper on the impact of AI on employment.
- *"AI and Jobs: The Role of Demand" by James Bessen*: Examines how AI affects labor demand and job creation.

3. **Websites and Online Portals**:

- *AI Now Institute*: Focuses on the social implications of AI, offering research and analysis.
- *Future of Life Institute*: Provides articles and resources on AI's impact on various aspects of life, including work.
- *McKinsey Global Institute*: Offers reports and insights on the economic impact of AI and automation.

4. **Conferences and Talks**:

- *TED Talks on Artificial Intelligence*: Features presentations from experts on AI and its societal impact.
- *NeurIPS (Conference on Neural Information Processing Systems)*: An annual event where the latest research in AI and machine learning is presented.
- *World Economic Forum Annual Meeting*: Discusses AI's role in shaping global economic trends.

5. **Documentaries and Video Series**:
   - *"Do You Trust This Computer?"*: Explores the promises and perils of AI.
   - *"The Age of AI" with Robert Downey Jr.*: A YouTube series examining how AI is changing the world.
   - *"AlphaGo"*: Chronicles the match between the Go champion and the AI program, AlphaGo, highlighting AI's capabilities.

6. **Educational Courses and Programs**:
   - *Coursera's "AI For Everyone" by Andrew Ng*: A course designed for non-technical people to understand AI.
   - *MIT OpenCourseWare on Artificial Intelligence*: Offers free course materials on AI from the Massachusetts Institute of Technology.
   - *edX's MicroMasters Program in Artificial Intelligence*: Provides an in-depth understanding of AI and its applications.

7. **Blogs and Podcasts**:
   - *"Lex Fridman Podcast"*: Features interviews with experts in AI and technology.
   - *"AI in Business"*: A podcast exploring AI's impact in the business world.
   - *"The Algorithm" by MIT Technology Review*: A newsletter that keeps readers informed on AI developments.

8. **Government and Policy Reports**:
   - *"Artificial Intelligence and Life in 2030" by Stanford University's One Hundred Year Study on Artificial Intelligence*: A comprehensive report on how AI will affect various aspects of life.
   - *European Commission's "Ethics Guidelines for Trustworthy AI"*: Provides guidelines for ethical AI development and use.

These resources offer a blend of theoretical knowledge, practical insights, and current research, making them invaluable for anyone interested in understanding the multifaceted implications of AI in the workforce.

References

15

References

Top of Form

- Abbate, J. (2000). *Inventing the Internet.* MIT Press.
- Accenture. (2016). Why Artificial Intelligence is the Future of Growth.
- Accenture. (2017). Harnessing the Potential of Public-Private Partnerships: The Catalytic Role of Multistakeholder Partnerships in the New Economy.
- Acemoglu, D., & Autor, D. (2011). Skills, Tasks and Technologies: Implications for Employment and Earnings. Handbook of Labor Economics.
- Acemoglu, D., & Autor, D. (2012). What Does Human Capital Do? A Review of Goldin and Katz's The Race between Education and Technology. Journal of Economic Literature.
- Acemoglu, D., & Restrepo, P. (2017). Robots and jobs: Evidence from US labor markets. Journal of Political Economy, 128(6), 2152-2186.
- Acemoglu, D., & Restrepo, P. (2018). The race between machine and man: Implications of technology for growth, factor shares, and employment. *American Economic Review.*

- Acemoglu, D., & Restrepo, P. (2019). Automation and new tasks: How technology displaces and reinstates labor. Journal of Economic Perspectives, 33(2), 3-30.
- Acemoglu, D., & Restrepo, P. (2020). Automation and new tasks: How technology displaces and reinstates labor. *Journal of Economic Perspectives.*
- Ackerman, K. (2016). The potential for urban agriculture in New York City: Growing capacity, food security, and green infrastructure. Urban Design Lab.
- Adam, A. (2005). Gender, Ethics and Information Technology. Palgrave Macmillan.
- Agarwal, R., Gao, G., DesRoches, C., & Jha, A. K. (2020). The digital transformation of healthcare: current status and the road ahead. Information Systems Research, 31(4), 1231-1259.
- Aghion, P., Jones, B. F., & Jones, C. I. (2017). Artificial Intelligence and Economic Growth. NBER Working Paper Series.
- Aghion, P., Jones, B. F., & Jones, C. I. (2019). Artificial Intelligence and Economic Growth. In Aghion, P., & Jones, B. F. (Eds.), The Economics of Artificial Intelligence: An Agenda. University of Chicago Press.
- Agrawal, A., Catalini, C., & Goldfarb, A. (2015). The Geography of Crowdfunding. National Bureau of Economic Research.
- Agrawal, A., et al. (2018). Prediction Machines: The Simple Economics of Artificial Intelligence. Harvard Business Review Press.
- Agrawal, A., Gans, J., & Goldfarb, A. (2017). What to Expect From Artificial Intelligence. *MIT Sloan Management Review.*
- Agrawal, A., Gans, J., & Goldfarb, A. (2018). *Prediction Machines: The Simple Economics of Artificial Intelligence.* Harvard Business Review Press.
- Agrawal, A., Gans, J., & Goldfarb, A. (2019). Prediction Machines: The Simple Economics of Artificial Intelligence. Harvard Business Review Press.

- Allen, I. E., & Seaman, J. (2014). Grade Change: Tracking Online Education in the United States. Babson Survey Research Group.
- Aloisi, A. (2015). Commoditized Workers: Case Study Research on Labour Law Issues Arising from a Set of 'On-Demand/Gig Economy' Platforms. Comparative Labor Law & Policy Journal.
- Aloisi, A., & De Stefano, V. (2021). Regulating Work in the Gig Economy: What Are the Options?. The Economic and Labour Relations Review.
- Amabile, T. (1996). Creativity in context. Westview Press.
- Amazon Annual Report. (2020). Annual Report 2020.
- Amazon. (2019). Amazon Pledges to Upskill 100,000 U.S. Employees for In-Demand Jobs by 2025.
- American Medical Association. (2020). The future of healthcare: a national survey of physicians.
- Anderson, J., & Rainie, L. (2017). The future of jobs and jobs training. Pew Research Center.
- Angwin, J., Larson, J., Mattu, S., & Kirchner, L. (2016). Machine bias. ProPublica.
- Arner, D. W., Barberis, J., & Buckley, R. P. (2017). FinTech and RegTech in a Nutshell, and the Future in a Sandbox. *Research in International Business and Finance*, 31, 207-233.
- Arntz, M., Gregory, T., & Zierahn, U. (2016). The Risk of Automation for Jobs in OECD Countries: A Comparative Analysis. *OECD Social, Employment and Migration Working Papers*, No. 189.
- Ashton, T. S. (1948). *The Industrial Revolution, 1760-1830.* Oxford University Press.
- AT&T Annual Report. (2020). Annual Report 2020.
- AT&T. (2016). AT&T Investing $1 Billion to Equip Workforce with Skills for Digital Future.
- Atkinson, A. B. (2015). *Inequality: What Can Be Done?* Harvard University Press.
- Atkinson, A. B., & Bourguignon, F. (Eds.). (2014). Handbook of Income Distribution. Elsevier.

- Atkinson, R. D., & Wu, J. (2017). False Alarmism: Technological Disruption and the U.S. Labor Market, 1850–2015. Information Technology and Innovation Foundation.
- Audretsch, D. B. (2015). Everything in its place: Entrepreneurship and the strategic management of cities, regions, and states. Oxford University Press.
- Audretsch, D. B., & Link, A. N. (2019). Entrepreneurship and innovation: Public policy frameworks. The Journal of Technology Transfer.
- Autor, D. H. (2015). Why are there still so many jobs? The history and future of workplace automation. Journal of Economic Perspectives, 29(3), 3-30.
- Autor, D. H. (2019). Work of the past, work of the future. AEA Papers and Proceedings, 109, 1-32.
- Autor, D. H., Levy, F., & Murnane, R. J. (2003). The Skill Content of Recent Technological Change: An Empirical Exploration. *Quarterly Journal of Economics*, 118(4), 1279-1333.
- Autor, D., & Dorn, D. (2013). The growth of low-skill service jobs and the polarization of the US labor market. *American Economic Review.*
- Autor, D., Dorn, D., & Hanson, G. (2015). Untangling trade and technology: Evidence from local labor markets. *The Economic Journal.*
- Autor, D., Mindell, D., & Reynolds, E. (2020). The Work of the Future: Building Better Jobs in an Age of Intelligent Machines. MIT Press.
- Back, A. L., Arnold, R. M., Baile, W. F., Fryer-Edwards, K. A., & Alexander, S. C. (2009). Approaching difficult communication tasks in oncology. CA: A Cancer Journal for Clinicians, 59(2), 113-126.
- Baker, R. S., & Inventado, P. S. (2014). Educational data mining and learning analytics. Springer.

- Baker, R. S., & Smith, L. (2019). Educational Data Mining and Learning Analytics. Cambridge University Press.
- Baker, S. (2021). Walmart's Education Program: Expanding Access and Opportunity. Walmart Newsroom.
- Baldwin, R. (2019). The globotics upheaval: Globalization, robotics, and the future of work. Oxford University Press.
- Bao, Y., Ke, B., Li, B., Lu, L., & Zhang, J. (2021). Artificial intelligence in finance. Annual Review of Financial Economics, 13.
- Barocas, S., Hardt, M., & Narayanan, A. (2019). Fairness and Abstraction in Sociotechnical Systems. ACM FAT* Conference.
- Bates, A. W. (2015). Teaching in a digital age: Guidelines for designing teaching and learning. Tony Bates Associates Ltd.
- Beauchamp, T. L., & Childress, J. F. (2012). Principles of biomedical ethics. Oxford University Press.
- Becker, K., & Sutcliffe, K. (2006). Managing the human side of medicine. *Journal of Healthcare Management.*
- Bell, S. (2010). Project-Based Learning for the 21st Century: Skills for the Future. The Clearing House: A Journal of Educational Strategies, Issues, and Ideas.
- Bennis, W. (2009). On becoming a leader. Basic Books.
- Berg, J. (2016). Income Security in the On-Demand Economy: Findings and Policy Lessons from a Survey of Crowdworkers. Comparative Labor Law & Policy Journal.
- Bergman, A. (2020). Bridging Academia and Industry: Ericsson's Collaborative Approach. AI & Education Review.
- Berman, S. J. (2012). Digital transformation: Opportunities to create new business models. Strategy & Leadership.
- Bersin, J. (2019). The Rise of the Corporate Academy: Salesforce Trailhead's Impact. Bersin by Deloitte.
- Bessen, J. E. (2015). Learning by Doing: The Real Connection between Innovation, Wages, and Wealth. Yale University Press.
- Bessen, J. E. (2016). Learning by Doing: The Real Connection between Innovation, Wages, and Wealth. Yale University Press.

- Bessen, J. E. (2019). AI and Jobs: The role of demand. *National Bureau of Economic Research*, Working Paper 24235.
- Bhatia, M., Sankaranarayanan, S., Poon, S. H., & Feldman, D. (2019). The role of artificial intelligence in decision-making. *Journal of Risk Management in Financial Institutions*, 12(3), 194-207.
- Bhatt, A. (2019). Digital health innovation: A toolkit to navigate from concept to clinical testing. JMIR Publications.
- Bholat, D., Lastra, R. M., Markose, S., Miglionico, A., & Sen, K. (2018). Machine learning in financial services: Changing the rules of the game. *Bank of England.*
- Binkley, M., et al. (2012). Defining twenty-first century skills. In Griffin, P., McGaw, B., & Care, E. (Eds.), Assessment and Teaching of 21st Century Skills. Springer.
- Bivens, J., García, E., Gould, E., Shierholz, H., & Wilson, V. (2016). It's Time for an Ambitious National Investment in America's Children: Investments in Early Childhood Care and Education Would Have Enormous Benefits for Children, Families, Society, and the Economy. Economic Policy Institute.
- Blease, C., Kaptchuk, T. J., Bernstein, M. H., Mandl, K. D., Halamka, J. D., & DesRoches, C. M. (2019). Artificial intelligence and the future of primary care: exploratory qualitative study of UK general practitioners' views. Journal of Medical Internet Research, 21(3), e12802.
- Boden, M. A. (2004). The creative mind: Myths and mechanisms. Routledge.
- Boserup, E. (1965). The Conditions of Agricultural Growth: The Economics of Agrarian Change under Population Pressure. Aldine.
- Bostrom, N. (2014). Superintelligence: Paths, Dangers, Strategies. Oxford University Press.
- Bostrom, N., & Dafoe, A. (2021). *Strategic Implications of Openness in AI Development.* Global Policy, 12(3), 245-256.

- Bostrom, N., & Yudkowsky, E. (2014). The ethics of artificial intelligence. In The Cambridge Handbook of Artificial Intelligence. Cambridge University Press.
- Boud, D., & Garrick, J. (1999). Understanding Learning at Work. Routledge.
- Boud, D., Keogh, R., & Walker, D. (1985). Reflection: Turning Experience into Learning. Kogan Page.
- Boudreau, J. & Ziskin, I. (2019). Preparing for the Future of Work: Challenges and Opportunities. Journal of Business Strategy.
- Bourdieu, P. (1993). The field of cultural production: Essays on art and literature. Columbia University Press.
- Bower, M. (2017). Design of Technology-Enhanced Learning: Integrating Research and Practice. Emerald Publishing Limited.
- Boyd, D., & Crawford, K. (2012). Critical questions for big data: Provocations for a cultural, technological, and scholarly phenomenon. Information, Communication & Society, 15(5), 662-679.
- Bradley, R., & Lawrence, N. (2019). Artificial intelligence and the environment in Australia and New Zealand. *AI & SOCIETY*, 34(3), 495-509.
- Bregman, R. (2017). Utopia for Realists: How We Can Build the Ideal World. Little, Brown, and Company.
- Bresnahan, T. F., & Trajtenberg, M. (1995). General purpose technologies 'Engines of growth'? Journal of Econometrics, 65(1), 83-108.
- Bridgstock, R. (2009). The graduate attributes we've overlooked: Enhancing graduate employability through career management skills. Higher Education Research & Development.
- Brooks, R. A. (2002). *Robot: The Future of Flesh and Machines.* Penguin Books.
- Brougham, D., & Haar, J. (2018). Smart technology, artificial intelligence, robotics, and algorithms (STARA): Employees'

perceptions of our future workplace. *Journal of Management & Organization, 24*(2), 239-257.

- Brown, M. (2022). Lifelong Learning in the Digital Age: The Trailhead Approach. Journal of Online Education.
- Brown, M. E., & Treviño, L. K. (2006). Ethical leadership: A review and future directions. The Leadership Quarterly, 17(6), 595-616.
- Brown, P., & Lauder, H. (2021). The New Collar Workforce: Reskilling in the Digital Age. Journal of Education and Work.
- Brynjolfsson, E., & McAfee, A. (2014). *The Second Machine Age: Work, Progress, and Prosperity in a Time of Brilliant Technologies.* W. W. Norton & Company.
- Brynjolfsson, E., & McAfee, A. (2016). *The Second Machine Age: Work, Progress, and Prosperity in a Time of Brilliant Technologies.* W.W. Norton & Company.
- Brynjolfsson, E., Mitchell, T., & Rock, D. (2018). What can machines learn, and what does it mean for occupations and the economy? AEA Papers and Proceedings, 108, 43-47.
- Brynjolfsson, E., Mitchell, T., & Rock, D. (2019). What can machines learn, and what does it mean for occupations and the economy? AEA Papers and Proceedings, 109, 43-47.
- Brynjolfsson, E., Rock, D., & Syverson, C. (2018). Artificial intelligence and the modern productivity paradox: A clash of expectations and statistics. *NBER Working Paper Series.*
- Bryson, J. J. (2018). AI & Global Governance: No one should trust AI. United Nations University Centre for Policy Research.
- Buckingham Shum, S., & Ferguson, R. (2012). Social Learning Analytics. Educational Technology & Society.
- Bughin, J., Chui, M., & Manyika, J. (2018). Notes from the AI frontier: Applications and value of deep learning. *McKinsey Global Institute.*
- Bughin, J., et al. (2018). Skill shift: Automation and the future of the workforce. McKinsey Global Institute.

- Buolamwini, J., & Gebru, T. (2018). Gender shades: Intersectional accuracy disparities in commercial gender classification. *Proceedings of the 1st Conference on Fairness, Accountability and Transparency.*
- Bureau of Labor Statistics. (2020). Occupational Outlook Handbook: Information Security Analysts.
- Burgsteiner, H., Kröll, M., Leopold-Wildburger, U., & Steinhardt, G. (2016). Learning Analytics for Educational Design in MOOCs and Flipped Classrooms. Studies in Systems, Decision and Control.
- Burrell, J. (2016). How the machine 'thinks': Understanding opacity in machine learning algorithms. Big Data & Society.
- Burtch, G., Carnahan, S., & Greenwood, B. N. (2018). Can You Gig It? An Empirical Examination of the Gig-Economy and Entrepreneurial Activity. Management Science.
- Byram, M. (2008). From foreign language education to education for intercultural citizenship. Multilingual Matters.
- Calo, R. (2015). Robotics and the Lessons of Cyberlaw. California Law Review.
- Campbell-Kelly, M., & Aspray, W. (2004). *Computer: A History of the Information Machine.* Westview Press.
- Capgemini. (2019). The Digital Talent Gap: Are Companies Doing Enough?
- Carnegie Mellon University. (2018). Interdisciplinary Programs. CMU.
- Case, A., & Deaton, A. (2020). Deaths of despair and the future of capitalism. Princeton University Press.
- Castells, M. (2001). The Internet Galaxy: Reflections on the Internet, Business, and Society. Oxford University Press.
- Castro, D., McLaughlin, M., & Chivot, E. (2019). How countries are pursuing an AI advantage: Insights from 25 countries' strategies. *Center for Data Innovation.*

- Cath, C., et al. (2018). Artificial intelligence and the 'good society': the US, EU, and UK approach. Science and Engineering Ethics.
- Chen, L., & Sheldon, M. (2016). Dynamic Pricing in a Labor Market: Surge Pricing and Flexible Work on the Uber Platform. Proceedings of the 2016 ACM Conference on Economics and Computation.
- Chen, M. K., Chevalier, J. A., Rossi, P. E., & Oehlsen, E. (2016). The Value of Flexible Work: Evidence from Uber Drivers. National Bureau of Economic Research.
- Chen, M. K., et al. (2019). The Value of Flexible Work: Evidence from Uber Drivers. Journal of Political Economy.
- Chetty, R., Hendren, N., Jones, M. R., & Porter, S. R. (2020). The Opportunity Atlas: Mapping the Childhood Roots of Social Mobility. NBER Working Paper No. 25147.
- Chetty, R., Hendren, N., Kline, P., & Saez, E. (2016). The Life-cycle of Scholarly Articles Across Fields of Research. Quarterly Journal of Economics.
- Choi, S. P. (2020). AI strategy in South Korea. *Science and Public Policy*, 47(3), 383-392.
- Choi, T. M., Wallace, S. W., & Wang, Y. (2018). Big data analytics in operations management. Production and Operations Management, 27(10), 1868-1883.
- Chui, M., Manyika, J., & Miremadi, M. (2016). Where machines could replace humans—and where they can't (yet). *McKinsey Quarterly*.
- Cohen, B., & Hopkins, D. (2019). Autonomous vehicles and the future of urban tourism. Annals of Tourism Research, 74, 33-42.
- Cohen, P., & Sundararajan, A. (2015). Self-Regulation and In-novation in the Peer-to-Peer Sharing Economy. University of Chicago Law Review Dialogue.
- Cohen, S. S., & Zysman, J. (1987). *Manufacturing Matters: The Myth of the Post-Industrial Economy*. Basic Books.

- Conley, D. T. (2019). A New Era for Educational Assessment. Students at the Center: Deeper Learning Research Series.
- Cook, T. D. (2020). The Safety and Security of Ride-Hailing Services: A Comparative Analysis. Transport Policy.
- Corbett, C., & Hill, C. (2015). Solving the Equation: The Variables for Women's Success in Engineering and Computing. American Association of University Women.
- Corey, G., Corey, M. S., & Callanan, P. (2014). Issues and ethics in the helping professions. Brooks/Cole, Cengage Learning.
- Costa-Font, M., Gil, J. M., & Traill, W. B. (2008). Consumer acceptance, valuation of and attitudes towards genetically modified food: Review and implications for food policy. Food Policy, 33(2), 99-111.
- Crafts, N. (2004). Steam as a General Purpose Technology: A Growth Accounting Perspective. The Economic Journal, 114(495), 338-351.
- Cramer, J., & Krueger, A. B. (2016). Disruptive Change in the Taxi Business: The Case of Uber. American Economic Review.
- Crawford, K., & Calo, R. (2016). There is a blind spot in AI research. Nature, 538(7625), 311-313.
- Csikszentmihalyi, M. (1990). Flow: The Psychology of Optimal Experience. Harper & Row.
- Csikszentmihalyi, M. (1996). Creativity: Flow and the psychology of discovery and invention. HarperCollins.
- da Silva, L. F., Neto, P. A., & Lima, C. R. (2021). The role of AI in sustainable resource management: The case of Brazil. *Journal of Environmental Management*, 281, 111827.
- Daniel, J. (2012). Making Sense of MOOCs: Musings in a Maze of Myth, Paradox and Possibility. Journal of Interactive Media in Education.
- Darling-Hammond, L., Flook, L., Cook-Harvey, C., Barron, B., & Osher, D. (2020). Implications for educational practice of the science of learning and development. Applied Developmental Science.

- Daugherty, P. R., & Wilson, H. J. (2018). Human + Machine: Reimagining Work in the Age of AI. Harvard Business Review Press.
- Dauth, W., Findeisen, S., Südekum, J., & Woessner, N. (2017). German robots - The impact of industrial robots on workers. IAB Discussion Paper.
- Davenport, T. H., & Kalakota, R. (2019). The potential for artificial intelligence in healthcare. Future Healthcare Journal, 6(2), 94-98.
- Davenport, T. H., & Kirby, J. (2016). Only humans need apply: Winners and losers in the age of smart machines. Harper Business.
- Davenport, T. H., & Patil, D. J. (2012). Data scientist: The sexiest job of the 21st century. Harvard Business Review, 90(10), 70-76.
- Davenport, T. H., & Ronanki, R. (2018). Artificial intelligence for the real world. *Harvard Business Review.*
- Davenport, T. H., Guha, A., Grewal, D., & Bressgott, T. (2020). How artificial intelligence will change the future of marketing. *Journal of the Academy of Marketing Science.*
- David, P. A. (1990). The Dynamo and the Computer: An Historical Perspective on the Modern Productivity Paradox. American Economic Review, 80(2), 355-361.
- Davies, R., Dean, D., & Ball, N. (2011). Flipping the classroom and instructional technology integration in a college-level information systems spreadsheet course. Frontiers in Education.
- Davis, G. F. (2016). The Vanishing American Corporation: Navigating the Hazards of a New Economy. Berrett-Koehler Publishers.
- Davis, K. (2020). Diversity and Inclusion in Tech Education: The P-TECH Approach. Tech Diversity Magazine.
- De Stefano, V. (2015). The Rise of the "Just-in-Time Workforce": On-Demand Work, Crowdwork, and Labor Protection in the "Gig-Economy". Comparative Labor Law & Policy Journal.

- De Stefano, V. (2016). The Rise of the "Just-in-Time Work-force": On-Demand Work, Crowdwork, and Labor Protection in the "Gig-Economy". Comparative Labor Law & De
- De Stefano, V. (2018). The Rise of the Just-in-Time Workforce: On-Demand Work, Crowdwork, and Labor Protection in the Gig-Economy. Comparative Labor Law & Policy Journal.
- De Stefano, V. (2020). The Rise of the Just-in-Time Workforce: On-Demand Work, Crowdwork, and Labor Protection in the Gig Economy. Comparative Labor Law & Policy Journal.
- Deardorff, D. K. (2006). The identification and assessment of intercultural competence as a student outcome of internationalization. Journal of Studies in International Education.
- Deci, E. L., & Ryan, R. M. (2000). The "What" and "Why" of Goal Pursuits: Human Needs and the Self-Determination of Behavior. Psychological Inquiry.
- Deming, D. J. (2017). The Growing Importance of Social Skills in the Labor Market. The Quarterly Journal of Economics.
- Deng, L., & Joshi, K. D. (2016). Why Do Freelancers Freelance? An Empirical Analysis from the Perspective of Behavioral Economics. IEEE Transactions on Engineering Management.
- Deng, L., Joshi, K., & Zhang, J. (2018). AI Challenges in Human-Robot Cognitive Collaboration. Cognitive Computation.
- Deng, X., Joshi, K. D., & Galliers, R. D. (2020). The Duality of Technology: Rethinking the Concept of Technology in Organizations. Journal of Management Studies.
- Densen, P. (2011). Challenges and opportunities facing medical education. Transactions of the American Clinical and Climatological Association, 122, 48-58.
- Despommier, D. (2010). The Vertical Farm: Feeding the World in the 21st Century. Thomas Dunne Books.
- Dewey, J. (1934). Art as experience. Minton, Balch & Company.
- Dincer, I., & Rosen, M. A. (2020). Sustainable Energy Technologies and Assessments, 40, 100760.

- Dossey, B. M., & Keegan, L. (2013). Holistic nursing: A handbook for practice. Jones & Bartlett Learning.
- Duggan, J., Sherman, U., Carbery, R., & McDonnell, A. (2020). Algorithmic Management and App-work in the Gig Economy: A Research Agenda for Employment Relations and HRM. Human Resource Management Journal.
- Durlak, J. A., Weissberg, R. P., Dymnicki, A. B., Taylor, R. D., & Schellinger, K. B. (2011). The impact of enhancing students' social and emotional learning: A meta-analysis of school-based universal interventions. *Child Development, 82*(1), 405-432.
- Dweck, C. S. (2007). Self-Theories: Their Role in Motivation, Personality, and Development. Psychology Press.
- Eckhardt, A., et al. (2018). The transformation of people, processes, and IT in e-Recruiting: Insights from an eight-year case study of a German media corporation. Employee Relations.
- Eisner, E. W. (2002). The Arts and the Creation of Mind. Yale University Press.
- Elias, M. J., Zins, J. E., Weissberg, R. P., Frey, K. S., Greenberg, M. T., Haynes, N. M., ... & Shriver, T. P. (1997). Promoting social and emotional learning: Guidelines for educators. ASCD.
- Eraut, M. (2004). Informal learning in the workplace. Studies in Continuing Education.
- Erickson, T. J. (2019). Getting to 'Us': Leveraging Multiculturalism for a Global Workforce. Harvard Business Review.
- Ericsson Ethics Board. (2021). Responsible AI Development at Ericsson.
- Ericsson Global. (2019). Annual Report: Fostering a Global AI Workforce.
- Ericsson HR. (2022). Training for the Future: Evaluating Ericsson's AI Program.
- Ericsson. (2020). Launching the Global AI Accelerator Program.
- Esteva, A., Kuprel, B., Novoa, R. A., Ko, J., Swetter, S. M., Blau, H. M., & Thrun, S. (2019). Dermatologist-level classification

of skin cancer with deep neural networks. *Nature,* 542(7639), 115–118.

- Esteva, A., Robicquet, A., Ramsundar, B., Kuleshov, V., DePristo, M., Chou, K., ... & Dean, J. (2019). A guide to deep learning in healthcare. Nature medicine, 25(1), 24-29.
- Eubanks, V. (2018). Automating Inequality: How High-Tech Tools Profile, Police, and Punish the Poor. St. Martin's Press.
- Eurofound (2020). Telework and ICT-based Mobile Work: Flexible Working in the Digital Age. European Foundation for the Improvement of Living and Working Conditions.
- European Commission. (2020). Digital Economy and Society Index (DESI) 2020.
- European Commission. (2021). *AI and the European Union: Balancing investment and regulation.* Retrieved from [European Commission website].
- European Commission's High-Level Expert Group on Artificial Intelligence. (2019). Ethics Guidelines for Trustworthy AI. European Commission.
- Fadel, C. (2016). Preparing Our Youth for an Inclusive and Sustainable World: The OECD PISA Global Competence Framework.
- Fallowfield, L., & Jenkins, V. (2004). Communicating sad, bad, and difficult news in medicine. Lancet, 363(9405), 312-319.
- Farrell, D., & Greig, F. (2016). Paychecks, Paydays, and the Online Platform Economy: Big Data on Income Volatility. JPMorgan Chase & Co. Institute.
- Federico, G. (2005). Feeding the World: An Economic History of Agriculture, 1800-2000. Princeton University Press.
- Feine, J., Morana, S., & Maedche, A. (2019). A Taxonomy of Social Cues for Conversational Agents. *International Journal of Human-Computer Studies,* 132, 138-161.

- Ferguson, R. (2019). Analytics and the use of student data in higher education: Promise and challenge. Journal of Interactive Media in Education.
- Ferguson, R., & Clow, D. (2017). Where are we now? And where are we going? In E. Langran & J. Borup (Eds.), Proceedings of Society for Information Technology & Teacher Education International Conference.
- Fernandez, A. (2019). Corporate Upskilling in the Digital Age: Walmart's Approach. Business Education Journal.
- Ferrari, E. (2018). The Rise of AI in the Workplace: A Guide to the Future of Work. Journal of Business and Technology.
- Ferrari, E., Pianesi, F., & Zancanaro, M. (2018). Human-Computer Interaction and Machine Learning. In Proceedings of the International Conference on Advanced Visual Interfaces.
- Fiala, B., et al. (2019). Interdisciplinary: Challenges and opportunities in higher education. AI & Society.
- Finnish Government. (2019). Finland offers AI training to every EU citizen.
- Fleming, N. D., & Mills, C. (1992). Not Another Inventory, Rather a Catalyst for Reflection. To Improve the Academy.
- Fleming, P. (2020). Dark Side of the Gig Economy: Workers' Perspectives and the Service Economy. Journal of Services Marketing.
- Florida, R. (2012). The rise of the creative class, revisited. Basic Books.
- Floridi, L., Cowls, J., Beltrametti, M., Chatila, R., Chazerand, P., Dignum, V., ... & Schafer, B. (2018). AI4People—An Ethical Framework for a Good AI Society: Opportunities, Risks, Principles, and Recommendations. Minds and Machines, 28(4), 689-707.
- Food and Agriculture Organization of the United Nations. (2019). The State of Food and Agriculture 2019. Moving forward on food loss and waste reduction.

- Ford, M. (2015). *Rise of the Robots: Technology and the Threat of a Jobless Future*. Basic Books.
- Foreman, J. W. (2013). Data Smart: Using Data Science to Transform Information into Insight. Wiley.
- Forsythe, E., Kahn, L. B., Lange, F., & Wiczer, D. (2020). Labor demand in the time of COVID-19: Evidence from vacancy postings and UI claims. Journal of Public Economics, 189, 104238.
- Fosso Wamba, S., & Queiroz, M. M. (2020). Blockchain in the Supply Chain: Key Challenges and Potential Benefits. International Journal of Information Management, 52, 102014.
- Fosway Group. (2018). Digital Learning Realities.
- Fountaine, T., McCarthy, B., & Saleh, T. (2019). Building the AI-powered organization. Harvard Business Review.
- Frank, J. D., & Frank, J. B. (1991). Persuasion and healing: A comparative study of psychotherapy. Johns Hopkins University Press.
- Frank, M., Roehrig, P., & Pring, B. (2019). What to do when machines do everything: How to get ahead in a world of AI, algorithms, bots, and big data. Wiley.
- Frenken, K., & Schor, J. (2017). Putting the Sharing Economy into Perspective. Environmental Innovation and Societal Transitions.
- Frey, C. B. (2019). *The Technology Trap: Capital, Labor, and Power in the Age of Automation*. Princeton University Press.
- Frey, C. B., & Osborne, M. A. (2017). The future of employment: How susceptible are jobs to computerization? *Technological Forecasting and Social Change*, 114, 254-280.
- Friedman, T. L. (2005). The World Is Flat: A Brief History of the Twenty-first Century. Farrar, Straus and Giroux.
- Furman, J., & Seamans, R. (2019). AI and the Economy. Innovation Policy and the Economy, 19(1), 161-191.
- Furnell, S. (2014). Making security usable: Are things improving?. Computers & Security, 43, 1-2.

- G20 Summit. (2019). G20 leaders' declaration. G20 Information Centre.
- Garcia, E. (2022). Corporate Social Responsibility in Action: Walmart's Education Initiatives. Journal of Business Ethics.
- Gardner, H. (1982). Art, mind, and brain: A cognitive approach to creativity. Basic Books.
- Gardner, H. (1983). Frames of mind: The theory of multiple intelligences. Basic Books.
- Gartner. (2018). Gartner Predicts 2018: CRM Customer Service and Support.
- Gimpel, H., et al. (2018). AI-based Services for Airbnb Hosts and Guests. Electronic Markets.
- Gnewuch, U., Morana, S., Maedche, A., & Weinhardt, C. (2017). Towards Designing Cooperative and Social Conversational Agents for Customer Service. *Proceedings of the 38th International Conference on Information Systems (ICIS 2017)*.
- Godfray, H. C. J., Beddington, J. R., Crute, I. R., Haddad, L., Lawrence, D., Muir, J. F., ... & Toulmin, C. (2010). Food security: The challenge of feeding 9 billion people. Science, 327(5967), 812-818.
- Golden, T. D. (2018). The Role of Telework in Work-Family Conflict. Journal of Vocational Behavior.
- Goldin, C. (2014). A Grand Gender Convergence: Its Last Chapter. *American Economic Review*.
- Goldin, C., & Katz, L. F. (2008). The race between education and technology. Belknap Press.
- Goleman, D. (1995). Emotional Intelligence. Bantam Books.
- Goleman, D. (2017). Emotional intelligence: Why it can matter more than IQ. Bantam Books.
- Gombrich, E. H. (1960). Art and illusion: A study in the psychology of pictorial representation. Phaidon.
- Gombrich, E. H. (2016). The story of art. Phaidon Press.

- Google. (2019). Grow with Google: Our commitment to helping Americans grow their skills, careers, and businesses.
- Google. (2020). Impact Report: Google IT Support Professional Certificate.
- Goos, M., & Manning, A. (2007). Lousy and lovely jobs: The rising polarization of work in Britain. The Review of Economics and Statistics, 89(1), 118-133.
- Goos, M., & Manning, A. (2007). Lousy and lovely jobs: The rising polarization of work in Britain. The Review of Economics and Statistics, 89(1), 118-133.
- Goos, M., Manning, A., & Salomons, A. (2014). Explaining job polarization: Routine-biased technological change and offshoring. American Economic Review, 104(8), 2509-26.
- Graef, I., Husovec, M., & Purtova, N. (2018). Data Protection, Data Access, and Antitrust Law. Journal of Competition Law & Economics.
- Transfer: European Review of Labour and Research.
- Graham, M., Hjorth, I., & Lehdonvirta, V. (2017). Digital Labour and Development: Impacts of Global Digital Labour Platforms and the Gig Economy on Worker Livelihoods. Transfer: European Review of Labour and Research.
- Gray, M. L., & Suri, S. (2019). Ghost Work: How to Stop Silicon Valley from Building a New Global Underclass. Houghton Mifflin Harcourt.
- Green, M. F., & Olson, C. (2008). Internationalizing the campus: A user's guide. American Council on Education.
- Greene, J. D., Sommerville, R. B., Nystrom, L. E., Darley, J. M., & Cohen, J. D. (2019). An fMRI investigation of emotional engagement in moral judgment. *Science*.
- Greenwood, B. N., & Wattal, S. (2017). Show Me the Way to Go Home: An Empirical Investigation of Ride-Sharing and Alcohol-Related Motor Vehicle Fatalities. MIS Quarterly.

- Grewal, D., Noble, S. M., Roggeveen, A. L., & Nordfält, J. (2020). The future of in-store technology. *Journal of the Academy of Marketing Science*, 48(1), 96-113.
- Grewal, D., Roggeveen, A. L., & Nordfält, J. (2020). The future of retailing. Journal of Retailing, 96(1), 1-6.
- Grow with Google. (2018). IT Support Professional Certificate.
- Gupta, P. (2020). Trailhead's Commitment to Diversity and Inclusion. Salesforce Blog.
- Gustavsson, J., Cederberg, C., & Sonesson, U. (2011). Global food losses and food waste. FAO.
- Guttentag, D. (2015). Airbnb: Disruptive Innovation and the Rise of an Informal Tourism Accommodation Sector. Current Issues in Tourism.
- Hadnagy, C., & Fincher, M. (2015). Phishing Dark Waters: The Offensive and Defensive Sides of Malicious Emails. Wiley.
- Hagel, J., Brown, J. S., & Davison, L. (2016). Navigating a shifting landscape: Capturing value in the evolving mobility ecosystem. Deloitte Insights.
- Hagel, J., et al. (2016). Navigating a shifting landscape: Capturing value in the evolving mobility ecosystem. Deloitte University Press.
- Hajkowicz, S. A., Reeson, A., Rudd, L., Bratanova, A., Hodgers, L., Mason, C., & Boughen, N. (2019). Tomorrow's Digitally Enabled Workforce: Megatrends and scenarios for jobs and employment in Australia over the coming twenty years. *CSIRO.*
- Hall, J. V., & Krueger, A. B. (2018). An Analysis of the Labor Market for Uber's Driver-Partners in the United States. Industrial and Labor Relations Review.
- Hall, J. V., & Krueger, A. B. (2018). An Analysis of the Labor Market for Uber's Driver-Partners in the United States. Industrial and Labor Relations Review.
- Hall, S. (1976). Cultural studies: Two paradigms. Media, Culture & Society, 2(1), 57-72.

- Halpern, J. (2001). From detached concern to empathy: Humanizing medical practice. Oxford University Press.
- Hamari, J., Koivisto, J., & Sarsa, H. (2014). Does Gamification Work? – A Literature Review of Empirical Studies on Gamification. Proceedings of the 47th Hawaii International Conference on System Sciences.
- Hanson, L. (2010). Global citizenship, global health, and the internationalization of curriculum: A study of transformative potential. Journal of Studies in International Education.
- Harari, Y. N. (2016). Homo Deus: A Brief History of Tomorrow. *Harvill Secker.*
- Harris, S. D., & Krueger, A. B. (2015). A Proposal for Modernizing Labor Laws for Twenty-First-Century Work: The "Independent Worker". The Hamilton Project.
- Hase, S., & Kenyon, C. (2007). Heutagogy: A child of complexity theory. Complicity: An International Journal of Complexity and Education.
- Hattie, J. (2009). Visible learning: A synthesis of over 800 meta-analyses relating to achievement. Routledge.
- Heaven, W. D. (2019). Why deep-learning AIs are so easy to fool. Nature News.
- Heckman, J. J., & Mosso, S. (2014). The economics of human development and social mobility. Annual Review of Economics, 6, 689-733.
- Heeks, R. (2018). Using ICTs to achieve the United Nations' sustainable development goals: Findings from the field. *Information Technology for Development,* 24(3), 476-491.
- Helsdingen, A. S., & Van Gog, T. (2018). The Role of Reflection in Learning and Professional Development. Educational Psychology Review.
- Helsinki Times. (2019). Finland's AI Course for Citizens: A Global Example.

- Helyer, R. (2015). Learning through reflection: The critical role of reflection in work-based learning (WBL). Journal of Work-Applied Management.
- Henao, A., & Marshall, W. E. (2019). The Impact of Ride-Hailing on Vehicle Miles Traveled. Transportation.
- Hernández-de-Menéndez, M., et al. (2019). Educational trends for the development of soft skills in IT engineering students. Computers in Human Behavior.
- Hess, T., Matt, C., Benlian, A., & Wiesböck, F. (2016). Options for Formulating a Digital Transformation Strategy. MIS Quarterly Executive, 15(2), 123-139.
- Heun, M. K., Brockway, P. E., Taylor, P. G., Domingos, T., & Sakai, M. (2018). A physical supply-use table framework for energy analysis on the energy conversion chain. Applied Energy, 226, 1134-1162.
- Hill, D. R. (1974). *The Book of Knowledge of Ingenious Mechanical Devices*. Springer.
- Hillage, J., & Pollard, E. (1998). Employability: Developing a framework for policy analysis. Department for Education and Employment.
- Hobsbawm, E. J. (1996). *The Age of Revolution: Europe 1789-1848*. Vintage Books.
- Hobsbawm, E. J. (1999). *Industry and Empire: The Birth of the Industrial Revolution*. New Press.
- Hodge, G. A., & Greve, C. (2019). *On Public–Private Partnership Performance: A Contemporary Review*. Public Works Management & Policy, 24(1), 3-40.
- Honey, M. A., & Hilton, M. (2011). Learning Science Through Computer Games and Simulations. National Academies Press.
- Honey, M. A., & Kanter, D. E. (2013). Design, Make, Play: Growing the Next Generation of STEM Innovators. Routledge.

- Horton, J. J., & Zeckhauser, R. J. (2016). Owning, Using, and Renting: Some Simple Economics of the "Sharing Economy". National Bureau of Economic Research.
- Hounshell, D. A. (1984). *From the American System to Mass Production, 1800-1932: The Development of Manufacturing Technology in the United States.* Johns Hopkins University Press.
- Hounshell, D. A. (1984). *From the American System to Mass Production, 1800-1932: The Development of Manufacturing Technology in the United States.* Johns Hopkins University Press.
- Houston, S. (2018). Social work: A critical approach to practice. Sage.
- Howcroft, D., & Bergvall-Kåreborn, B. (2019). A Typology of Crowdwork Platforms. Work, Employment and Society.
- Howell, S. (2016). The Future of Work: The Rise of the Gig Economy. National Bureau of Economic Research.
- Hrastinski, S. (2008). Asynchronous and Synchronous E-Learning. EDUCAUSE Quarterly.
- Huang, J., et al. (2020). Lifelong learning in the AI era: A content analysis of trends, opportunities, and challenges. Computers & Education.
- Huang, M. H., & Rust, R. T. (2018). Artificial Intelligence in Service. Journal of Service Research.
- Huang, M. H., Rust, R. T., & Maksimovic, V. (2019). The AI revolution and the future of marketing. Journal of the Academy of Marketing Science.
- Huang, M.-H., & Rust, R. T. (2018). Artificial Intelligence in Service. *Journal of Service Research*, 21(2), 155-172.
- Hughes, C. (2018). Fair Shot: Rethinking Inequality and How We Earn. St. Martin's Press.
- Hughes, T. P. (1983). *Networks of Power: Electrification in Western Society, 1880-1930.* Johns Hopkins University Press.

- Hunter, B., White, G. P., & Godbey, G. C. (2006). What does it mean to be globally competent? Journal of Studies in International Education.
- Huws, U., et al. (2017). Work in the European Gig Economy – Research Results from the UK, Sweden, Germany, Austria, the Netherlands, Switzerland and Italy. Foundation for European Progressive Studies.
- Hwang, G. J. (2004). A study of multi-media annotation of Web-based materials. Computers & Education.
- Hwang, G. J. (2014). Definition, Framework and Research Issues of Smart Learning Environments - A Context-Aware Ubiquitous Learning Perspective. Smart Learning Environments.
- Hwang, G. J. (2020). Definition, framework and research issues of smart learning environments - a context-aware ubiquitous learning perspective. Smart Learning Environments.
- Hwang, G. J., & Lai, C. L. (2017). Facilitating and Bridging Out-of-Class and In-Class Learning: An Interactive E-Book-Based Flipped Learning Approach for Math Courses. Educational Technology & Society.
- IBM Annual Report. (2020). Building Workforce Skills at Scale.
- IBM. (2017). IBM's New Collar Initiatives.
- IBM. (2018). New Collar Jobs: Bridging the Skills Gap.
- IBM. (2019). P-TECH: Reinventing Education for the New Collar Workforce.
- IEEE. (2019). Ethically Aligned Design: A Vision for Prioritizing Human Well-being with Autonomous and Intelligent Systems. IEEE.
- International Renewable Energy Agency [IRENA]. (2018). Renewable Energy and Jobs – Annual Review 2018.
- International Telecommunication Union. (2021). AI Standards in the Telecom Industry: Ericsson's Contribution.
- IRENA. (2019). Renewable Energy and Jobs – Annual Review 2019. International Renewable Energy Agency.

- ITU. (2019). *Bridging the digital divide through public-private partnerships.* International Telecommunication Union. Retrieved from [ITU website].
- Ivanov, D., & Dolgui, A. (2020). Viability of intertwined supply networks: Extending the supply chain resilience angles towards survivability. A position paper motivated by COVID-19 outbreak. *International Journal of Production Research,* 58(10), 2904-2915.
- Ivanov, D., Dolgui, A., & Sokolov, B. (2019). The Impact of Digital Technology and Industry 4.0 on the Ripple Effect and Supply Chain Risk Analytics. International Journal of Production Research, 57(3), 829-846.
- Jackson, D. (2013). The contribution of work-integrated learning to undergraduate employability skill outcomes. Asia-Pacific Journal of Cooperative Education.
- Jacobson, M. Z., Delucchi, M. A., Cameron, M. A., & Frew, B. A. (2017). A 100% wind, water, sunlight (WWS) all-sector energy plan for 139 countries of the world. Joule, 1(1), 108-121.
- Jenkins, A., & Mostafa, T. (2015). The effects of learning on well-being for older adults in England. Ageing and Society.
- Jenkins, H. (2006). Convergence Culture: Where Old and New Media Collide. *NYU Press.*
- Jiang, F., Jiang, Y., Zhi, H., Dong, Y., Li, H., Ma, S., ... & Wang, Y. (2017). Artificial intelligence in healthcare: past, present and future. *Stroke and Vascular Neurology,* 2(4).
- 230-243.
- Jobin, A., Ienca, M., & Vayena, E. (2019). The Global Landscape of AI Ethics Guidelines. *Nature Machine Intelligence,* 1, 389–399.
- Johansson, L. (2021). Upskilling in AI: Ericsson's Strategic Approach. Technology Workforce Journal.
- Johnson, L. (2020). The Role of Corporate Training Programs in Digital Transformation. Corporate Learning Review.

- Johnson, L. (2022). Evolving digital education: Challenges in aligning curriculum with industry needs. Educational Technology Review.
- John-Steiner, V. (2000). Creative collaboration. Oxford University Press.
- Johnston, H., & Land-Kazlauskas, C. (2018). Organizing on-demand: Representation, voice, and collective bargaining in the gig economy. Conditions of Work and Employment Series, International Labour Office.
- Jones, A. (2000). The Antikythera Mechanism: The First Step towards the Future. *Ancient Astronomy and Celestial Divination*, 291-307.
- Jordan, M. I., & Mitchell, T. M. (2015). Machine learning: Trends, perspectives, and prospects. Science, 349(6245), 255-260.
- Jordan, M. I., & Mitchell, T. M. (2015). Machine learning: Trends, perspectives, and prospects. Science, 349(6245), 255-260.
- Kahn, W. A. (2018). The importance of human interaction in healthcare. *Healthcare Management Review.*
- Kahneman, D. (2011). Thinking, fast and slow. Farrar, Straus, and Giroux.
- Kalleberg, A. L., & Dunn, M. (2016). Good Jobs, Bad Jobs in the Gig Economy. Perspectives on Work.
- Kalleberg, A. L., & Vallas, S. P. (2018). Probing Precarious Work: Theory, Research, and Politics. Research in the Sociology of Work.
- Kalogirou, S. A. (2018). Artificial intelligence in renewable energy systems design and operation. Renewable Energy, 139, 1296-1302.
- Kaplan, A., & Haenlein, M. (2019). Siri, Siri, in my hand: Who's the fairest in the land? On the interpretations, illustrations, and implications of artificial intelligence. *Business Horizons*, 62(1), 15-25.

- Kaplan, J. (2016). Artificial Intelligence: What Everyone Needs to Know. Oxford University Press.
- Kaplan, J. (2020). The Promise and Perils of Corporate Training Programs in the Era of AI. Journal of Business Ethics.
- Kaplan, J., et al. (2020). The Gig Economy and the Future of Professional Work. McKinsey & Company.
- Kapoor, A. (2019). AI and the future workforce in India. *Journal of Business Ethics*, 160(4), 835-850.
- Kapoor, A., Lee, J., & Nair, H. (2018). The role of artificial intelligence in customer service. Customer Service Management Journal.
- Kässi, O., & Lehdonvirta, V. (2018). Online Labour Index: Measuring the Online Gig Economy for Policy and Research. Technological Forecasting and Social Change.
- Kates, R. W., Parris, T. M., & Leiserowitz, A. A. (2001). What is Sustainable Development? Goals, Indicators, Values, and Practice. Environment: Science and Policy for Sustainable Development.
- Katz, L. F., & Krueger, A. B. (2016). The Rise and Nature of Alternative Work Arrangements in the United States, 1995-2015. National Bureau of Economic Research.
- Katz, L. F., & Krueger, A. B. (2017). The Rise and Nature of Alternative Work Katz, L. F., & Krueger, A. B. (2019). The Rise and Nature of Alternative Work Arrangements in the United States. ILR Review.
- Katz, L. F., & Margo, R. A. (2014). Technical change and the relative demand for skilled labor: The United States in historical perspective. In Human Capital in History: The American Record (pp. 15-57). University of Chicago Press.
- Kenney, M., & Zysman, J. (2016). The Rise of the Platform Economy. Issues in Science and Technology.
- Kerr, S., & Walrath, R. (2017). Renewable energy's intermittency problem: Massive energy storage in a tiny footprint. Renewable and Sustainable Energy Reviews, 78, 934-945.

- Khandani, A. E., Kim, A. J., & Lo, A. W. (2010). Consumer credit-risk models via machine-learning algorithms. *Journal of Banking & Finance*, 34(11), 2767-2787.
- Khullar, D. (2018). The Role of Technology in Health Care Innovation: Insights from the COVID-19 Pandemic. Journal of Medical Internet Research, 20(5), e195.
- Kidder, R. M. (2009). How good people make tough choices: Resolving the dilemmas of ethical living. Harper.
- Kingsley, S. C., Gray, M. L., & Suri, S. (2015). Monopsony and the Crowd: Labor for Lemons? Industrial and Labor Relations Review.
- Knowles, M. S. (1975). Self-Directed Learning: A Guide for Learners and Teachers. Association Press.
- Knowles, M. S., Holton III, E. F., & Swanson, R. A. (2005). The Adult Learner: The Definitive Classic in Adult Education and Human Resource Development. Elsevier.
- Koivisto, J. (2020). Enhancing National Digital Literacy: The Finnish Model. Journal of Digital Education.
- Kolb, D. A. (2014). Experiential Learning: Experience as the Source of Learning and Development. Pearson Education.
- Kolk, A., Van Dolen, W., & Vock, M. (2020). Trickle effects of cross-sector social partnerships. *Journal of Business Ethics*, 162(3), 697-723.
- Korhonen, P. (2021). Building a Technologically Advanced Society: Finland's AI Strategy. Nordic Tech Review.
- Korinek, A., & Stiglitz, J. E. (2017). Artificial intelligence and its implications for income distribution and unemployment. *National Bureau of Economic Research*.
- Kowalski, R., & Swanson, D. (2017). Cybersecurity and Data Protection in a Globalized Work Environment. Information Systems Management.

- Krajcik, J. S., & Blumenfeld, P. C. (2006). Project-based learning. In Sawyer, R. K. (Ed.), The Cambridge Handbook of the Learning Sciences. Cambridge University Press.
- Kshetri, N. (2017). Big Data's Impact on Privacy, Security and Consumer Welfare. Telecommunications Policy.
- Kshetri, N. (2017). Cybersecurity: The Changing Nature of Cyber Threats Facing SMEs. Journal of International Affairs, 70(1), 77-90.
- Kuek, S. C., Paradi-Guilford, C., Fayomi, T., Imaizumi, S., Ipeirotis, P., Pina, P., & Singh, M. (2015). The Global Opportunity in Online Outsourcing. World Bank.
- Kuhn, T. S. (1962). The structure of scientific revolutions. University of Chicago Press.
- Kulik, J. A. (2016). An overview of the effectiveness of adaptive learning in education. International Journal of STEM Education.
- Kumar, A., Bezawada, R., Rishika, R., Janakiraman, R., & Kannan, P. K. (2019). From Social to Sale: The Effects of Firm-Generated Content in Social Media on Customer Behavior. Journal of Marketing.
- Laal, M., & Laal, M. (2012). Collaborative learning: What is it? Procedia - Social and Behavioral Sciences.
- Lam, S., & Liu, Y. (2018). When Does Demand Trump Supply? Ride-sharing and Congestion during Peak Hours. Transportation Research Part B: Methodological.
- Lancioni, G. E., et al. (2012). Assistive technology for people with severe/profound intellectual and multiple disabilities. In: Severe Intellectual Disability. Springer.
- Landes, D. S. (1969). *The Unbound Prometheus: Technological Change and Industrial Development in Western Europe from 1750 to the Present.* Cambridge University Press.
- Landes, D. S. (2003). *The Unbound Prometheus: Technological Change and Industrial Development in Western Europe from 1750 to the Present.* Cambridge University Press.

- Langlotz, C. P. (2019). Will artificial intelligence replace radiologists? *Radiology: Artificial Intelligence*, 1(3), e190058.
- Lee, I., et al. (2020). Big data analytics and AI in next-generation technology. IEEE Access.
- Lee, J., Bagheri, B., & Kao, H. A. (2019). A cyber-physical systems architecture for industry 4.0-based manufacturing systems. *Manufacturing Letters*, 15, 18-23.
- Lee, J., Kao, H. A., & Yang, S. (2014). Service innovation and smart analytics for Industry 4.0 and big data environment. *Procedia CIRP*, 16, 3-8.
- Lee, J., Kao, H. A., & Yang, S. (2018). Service innovation and smart analytics for industry 4.0 and big data environment. Procedia CIRP, 16, 3-8.
- Lee, K.-F. (2018). *AI Superpowers: China, Silicon Valley, and the New World Order*. Houghton Mifflin Harcourt.
- Lee, K.-F. (2020). Building an Inclusive AI Economy. Harvard Business Review.
- Lee, K.-F., & Fung, A. (2019). AI Superpowers: China, Silicon Valley, and the New World Order. Houghton Mifflin Harcourt.
- Lee, M. H., & Zhao, Z. (2020). Challenges in implementing lifelong learning systems: The case of Singapore. Journal of Lifelong Learning Society.
- Lee, M. K., et al. (2015). Working with Machines: The Impact of Algorithmic and Data-Driven Management on Human Workers. Proceedings of the 33rd Annual ACM Conference on Human Factors in Computing Systems.
- Lee, M. K., Kusbit, D., Metsky, E., & Dabbish, L. (2015). Working with Machines: The Impact of Algorithmic and Data-Driven Management on Human Workers. Proceedings of the 33rd Annual ACM Conference on Human Factors in Computing Systems.

- Leimeister, J. M. (2020). Seamless AI Integration in Work Processes: The Future of Work in the Gig Economy. Journal of Business Research.
- Leukfeldt, E. R., Holt, T. J., & Bernaards, C. A. (2017). Examining the predictors of cybercrime offending among youth. Journal of Contemporary Criminal Justice, 33(4), 368-388.
- Lewis, J. I. (2014). Green innovation in China: China's wind power industry and the global transition to a low-carbon economy. Columbia University Press.
- Li, C., & Ma, X. (2020). The role of artificial intelligence in learning process: A literature review. International Education Studies.
- Li, G., Weng, S., Liu, X., & Zhang, Y. (2019). A review of the applications of artificial intelligence in the management of renewable energy systems. Renewable and Sustainable Energy Reviews, 108, 320-331.
- Liakos, K. G., Busato, P., Moshou, D., Pearson, S., & Bochtis, D. (2018). Machine Learning in Agriculture: A Review. Sensors, 18(8), 2674.
- Libicki, M. C., Senty, D., & Pollak, J. (2014). Hackers Wanted: An Examination of the Cybersecurity Labor Market. RAND Corporation.
- Libicki, M. C., Senty, D., & Pollak, J. (2015). Cybersecurity in the Golden Age of Spying. IEEE Security & Privacy, 13(1), 14-21.
- Litman, T. (2020). Smart Transportation: New Technologies and Strategies to Improve Urban Travel. Victoria Transport Policy Institute.
- Litow, S. (2018). The P-TECH Model: Responding to Technological Change. Journal of Education and Training.
- Litow, S., & Hood, P. (2016). P-TECH: Pathways in Technology Early College High School.
- Lowrey, A. (2018). Give People Money: How a Universal Basic Income Would End Poverty, Revolutionize Work, and Remake the World. Crown.

- Luckin, R., et al. (2016). Intelligence Unleashed: An argument for AI in Education. Pearson.
- Luxton, D. D. (2020). Artificial intelligence in psychological practice: Current and future applications and implications. Professional Psychology: Research and Practice, 51(5), 466-474.
- Malhotra, A., & Van Alstyne, M. (2014). The Dark Side of the Sharing Economy... and How to Lighten It. Communications of the ACM.
- ManpowerGroup. (2020). The Skills Revolution Reboot: The 3Rs - Renew, Reskill, Redeploy.
- Mansilla, V. B., & Jackson, A. (2011). Educating for global competency: The value of multilingualism. Asia Society.
- Manyika, J., Chui, M., Miremadi, M., Bughin, J., George, K., Willmott, P., & Dewhurst, M. (2017). A Future That Works: Automation, Employment, and Productivity. McKinsey Global Institute.
- Manyika, J., et al. (2017). A Future that Works: Automation, Employment, and Productivity. McKinsey Global Institute.
- Manyika, J., Lund, S., Bughin, J., Robinson, K., Mischke, J., & Mahajan, D. (2017). Jobs lost, jobs gained: Workforce transitions in a time of automation. McKinsey Global Institute.
- Marcus, G. (2018). Deep Learning: A Critical Appraisal. arXiv preprint arXiv:1801.00631.
- Marr, B. (2015). Big Data: Using SMART Big Data, Analytics and Metrics To Make Better Decisions and Improve Performance. Wiley.
- Marr, B. (2016). *Data Strategy: How to Profit from a World of Big Data, Analytics and the Internet of Things*. Kogan Page.
- Martin, K. (2019). Ethical Implications and Accountability of Algorithms. Journal of Business Ethics.
- Mayer, J. D., & Salovey, P. (1997). What is emotional intelligence? In P. Salovey & D. Sluyter (Eds.), Emotional development and emotional intelligence: Educational implications. Basic Books.

- Mayer, J. D., Roberts, R. D., & Barsade, S. G. (2008). Human abilities: Emotional intelligence. Annual Review of Psychology, 59, 507-536.
- Mayer, R. E. (2014). Cognitive theory of multimedia learning. In The Cambridge Handbook of Multimedia Learning.
- Mayer-Schönberger, V., & Cukier, K. (2013). Big Data: A Revolution That Will Transform How We Live, Work, and Think. Eamon Dolan/Houghton Mifflin Harcourt.
- Mayer-Schönberger, V., & Cukier, K. (2013). Big Data: A Revolution That Will Transform How We Live, Work, and Think. John Murray.
- Mayor, A. (2018). *Gods and Robots: Myths, Machines, and Ancient Dreams of Technology*. Princeton University Press.
- Mazzucato, M. (2015). The Entrepreneurial State: Debunking Public vs. Private Sector Myths. Anthem Press.
- McKinsey & Company. (2018). Notes from the AI frontier: Insights from hundreds of use cases. Retrieved from [McKinsey Report].
- McKinsey & Company. (2019). Global AI Survey: AI proves its worth, but few scale impact. Retrieved from [McKinsey Report].
- McKinsey Global Institute. (2017). Jobs Lost, Jobs Gained: Workforce Transitions in a Time of Automation.
- McKinsey Global Institute. (2018). Skill shift: Automation and the future of the workforce. McKinsey & Company.
- McKinsey Global Institute. (2019). The Future of Work in America: People and Places, Today and Tomorrow.
- McSherry, W., & Ross, L. (2010). Spiritual assessment in healthcare practice. M&K Update Ltd.
- Means, B., et al. (2013). Evaluation of evidence-based practices in online learning: A meta-analysis and review of online learning studies. U.S. Department of Education.
- Mehta, M., Dahl, D. W., & Wattenhofer, R. (2018). How AI is changing the way companies are organized. Journal of Organization Design.

- Mehta, N. (2017). A New Culture of Learning: Cultivating the Imagination for a World of Constant Change. CreateSpace Independent Publishing Platform.
- Mehta, R., & Aguilera, E. (2018). A critical approach to humanizing pedagogies in online teaching and learning. The International Journal of Information and Learning Technology.
- Meister, J. (2017). Future Work Skills and Learning Strategies. Journal of Workplace Learning.
- Meister, J. (2020). The Future of Work: Preparing for Disruption. World Bank Group.
- Meuter, M. L., Bitner, M. J., Ostrom, A. L., & Brown, S. W. (2005). Choosing Among Alternative Service Delivery Modes: An Investigation of Customer Trial of Self-Service Technologies. *Journal of Marketing*, 69(2), 61-83.
- Microsoft. (2020). *Microsoft's AI initiative with community colleges.* Retrieved from [Microsoft website].
- Miller, R. (2017). Salesforce's Trailhead Creates a Fun Learning Environment. TechCrunch.
- Milligan, M., Ela, E., Hodge, B. M., Kirby, B., Lew, D., Clark, C., ... & Lynn, K. (2016). The role of power system flexibility in generation planning. IEEE Power and Energy Magazine, 14(6), 65-73.
- Ministry of Education, Singapore. (2016). Enhancing lifelong learning through SkillsFuture.
- Mittelstadt, B. (2019). AI ethics – too principled to fail? Journal of the British Academy, 7(s2), 17-24.
- Möhlmann, M. (2016). Collaborative Consumption: Determinants of Satisfaction and the Likelihood of Using a Sharing Economy Option Again. Journal of Consumer Behaviour.
- Möhlmann, M., & Zalmanson, L. (2017). Hands on the Wheel: Navigating Algorithmic Management and Uber Drivers' Autonomy. Proceedings of the International Conference on Information Systems (ICIS).

- Mokyr, J. (2002). The Gifts of Athena: Historical Origins of the Knowledge Economy. Princeton University Press.
- Mokyr, J., Vickers, C., & Ziebarth, N. L. (2015). The history of technological anxiety and the future of economic growth: Is this time different? Journal of Economic Perspectives, 29(3), 31-50.
- Moretti, E. (2012). The new geography of jobs. Houghton Mifflin Harcourt.
- Mushtaq, H. (2019). History of Ancient Engineering. *Technological Innovations in Ancient Times*, 1-15.
- National Science and Technology Council. (2016). Preparing for the Future of Artificial Intelligence.
- Needham, J. (1986). *Science and Civilization in China: Volume 4, Physics and Physical Technology, Part 2, Mechanical Engineering.* Cambridge University Press.
- Ng, P. T. (2018). Lifelong learning in Singapore: Policies and initiatives. International Journal of Lifelong Education.
- Niemi, M. (2021). AI for Everyone: The Finnish Strategy. AI & Society Journal.
- Nilsson, M. (2022). AI in Telecommunications: Ericsson's Competitive Edge. AI Business Review.
- Noble, S. U. (2018). Algorithms of Oppression: How Search Engines Reinforce Racism. *NYU Press*.
- Norcross, J. C., & Wampold, B. E. (2011). Evidence-based therapy relationships: Research conclusions and clinical practices. Psychotherapy, 48(1), 98-102.
- Nussbaum, M. C. (2010). Not for Profit: Why Democracy Needs the Humanities. Princeton University Press.
- Nygaard, I., & Wieczorek, A. J. (2018). Sustainable energy transitions in emerging economies: The formation of a palm oil biomass waste-to-energy niche in Malaysia 1990–2016. Environmental Innovation and Societal Transitions, 27, 10-22.
- O'Neil, C. (2016). Weapons of Math Destruction: How Big Data Increases Inequality and Threatens Democracy. Crown.

- Obermeyer, Z., Emanuel, E. J. (2016). Predicting the Future — Big Data, Machine Learning, and Clinical Medicine. *New England Journal of Medicine*, 375, 1216-1219.
- O'Dowd, R. (2018). Virtual exchange: Moving forward into the next decade. Computer Assisted Language Learning.
- OECD. (2021). Enhancing Access to and Sharing of Data: Reconciling Risks and Benefits for Data Re-use across Societies.
- OECD. (2021). *Regulatory frameworks for public-private partnerships: Ensuring good governance and societal benefit.* Organisation for Economic Co-operation and Development. Retrieved from [OECD website].
- Okeleke, K., & Suleyman, A. (2019). AI in Africa: A growing ecosystem. *GSMA Intelligence Report.*
- Oleson, J. P. (1984). Greek and Roman Mechanical Water-Lifting Devices: The History of a Technology. *Phoenix*, 38(2), 152-173.
- Oliver, B. (2015). Redefining graduate employability and work-integrated learning: Proposals for effective higher education in disrupted economies. Journal of Teaching and Learning for Graduate Employability.
- Oliver, B. (2019). Making micro-credentials work for learners, employers and providers. Deakin University.
- O'Neil, C. (2016). Weapons of Math Destruction: How Big Data Increases Inequality and Threatens Democracy. Crown.
- Organisation for Economic Co-operation and Development. (2019). OECD Employment Outlook 2019: The Future of Work. OECD Publishing.
- Orsini, F., Kahane, R., Nono-Womdim, R., & Gianquinto, G. (2013). Urban agriculture in the developing world: a review. Agronomy for Sustainable Development, 33(4), 695-720.
- Pager, D., & Shepherd, H. (2008). The Sociology of Discrimination: Racial Discrimination in Employment, Housing, Credit, and Consumer Markets. Annual Review of Sociology.

- Pane, J. F., et al. (2017). How does personalized learning affect student achievement? RAND Corporation.
- Pane, J. F., Steiner, E. D., Baird, M. D., & Hamilton, L. S. (2017). Informing Progress: Insights on Personalized Learning Implementation and Effects. RAND Corporation.
- Parker, K. (2019). AI and Mental Health: Digital Well-being in the Modern Workplace. Journal of Occupational Health Psychology.
- Partridge, M. D. (2015). The dueling models: NEG vs. amenity migration in explaining US engines of growth. *Papers in Regional Science.*
- Patrick, C. J., Peach, D., Pocknee, C., Webb, F., Fletcher, M., & Pretto, G. (2008). The WIL (Work Integrated Learning) report: A national scoping study. Australian Learning and Teaching Council (ALTC).
- Paul, R., & Elder, L. (2006). Critical thinking: The nature of critical and creative thought. Journal of Developmental Education, 30(2), 34.
- Pedersen, P. (1991). Multiculturalism as a generic approach to counseling. Journal of Counseling & Development, 70(1), 6-12.
- Pellegrino, E. D., & Thomasma, D. C. (1993). The virtues in medical practice. Oxford University Press.
- Perkins, D. (2014). Future wise: Educating our children for a changing world. Jossey-Bass.
- Piketty, T. (2020). *Capital and Ideology.* Harvard University Press.
- Pink, D. H. (2019). A whole new mind: Why right-brainers will rule the future. Riverhead Books.
- Pistorius, C. W. I., & Utterback, J. M. (1997). Multi-mode Interaction Among Technologies. *Research Policy,* 26(1), 67-84.
- Polonetsky, J., & Tene, O. (2014). Privacy and big data: Making ends meet. Stanford Law Review.
- Polonski, V. (2017). The impact of AI on cybersecurity. The World Economic Forum.

- Pomeranz, K. (2000). *The Great Divergence: China, Europe, and the Making of the Modern World Economy.* Princeton University Press.
- Porter, M. E. (2010). What is value in health care? New England Journal of Medicine, 363(26), 2477-2481.
- Porter, M. E., & Heppelmann, J. E. (2014). How smart, connected products are transforming competition. *Harvard Business Review.*
- Porter, M. E., & Stern, S. (2000). Measuring the 'Ideas' Production Function: Evidence from International Patent Output. NBER Working Paper No. 7891.
- Prassl, J. (2018). Humans as a Service: The Promise and Perils of Work in the Gig Economy. Oxford University Press.
- Pretty, J., Toulmin, C., & Williams, S. (2011). Sustainable Intensification in African Agriculture. International Journal of Agricultural Sustainability, 9(1), 5-24.
- Prinsloo, P., & Slade, S. (2017). Big data, higher education and learning analytics: Beyond justice, towards an ethics of care. Big Data and Cognitive Computing.
- Provost, F., & Fawcett, T. (2013). Data Science for Business: What You Need to Know about Data Mining and Data-Analytic Thinking. O'Reilly Media.
- P-TECH. (2019). The P-TECH Model: Public-Private Education Collaboration.
- P-TECH. (2021). P-TECH Global Impact Report.
- Putnam, R. D. (2000). Bowling Alone: The Collapse and Revival of American Community. Simon & Schuster.
- PwC. (2017). Global Artificial Intelligence Study: Exploiting the AI Revolution.
- PwC. (2017). Sizing the prize: What's the real value of AI for your business and how can you capitalise?
- PwC. (2018). Will robots really steal our jobs? An international analysis of the potential long term impact of automation.

- Qiu, J. L., Gregg, M., & Crawford, K. (2014). Circuits of Labour: A Labour Theory of the iPhone Era. TripleC.
- Rajan, A. (2020). AI and the future of work: Towards a sustainable model. *Journal of Business Ethics*, 165(3), 609-621.
- Reimers, F. M., & Chung, C. K. (2016). Teaching and learning for the twenty-first century: Educational goals, policies, and curricula from six nations. Harvard Education Press.
- Renda, A. (2019). Artificial Intelligence: Ethics, governance and policy challenges. *Centre for European Policy Studies*.
- Renner, M., Sweeney, S., & Kubit, J. (2008). Green Jobs: Towards Decent Work in a Sustainable, Low-Carbon World. United Nations Environment Programme.
- Ribble, M. (2015). Digital citizenship in schools: Nine elements all students should know. International Society for Technology in Education.
- Richards, N. M., & King, J. H. (2014). Big Data Ethics. *Wake Forest Law Review*, 49, 393-432.
- Rifkin, J. (2011). *The Third Industrial Revolution: How Lateral Power is Transforming Energy, the Economy, and the World*. Palgrave Macmillan.
- Rifkin, J. (2014). The Zero Marginal Cost Society: The Internet of Things, the Collaborative Commons, and the Eclipse of Capitalism. Palgrave Macmillan.
- Riordan, M., & Hoddeson, L. (1997). *Crystal Fire: The Invention of the Transistor and the Birth of the Information Age*. W. W. Norton & Company.
- Rios, F. & Siveroni, L. (2012). Collaborative problem-solving and the development of cultural competence. Intercultural Education.
- Robertson, I. T. (2020). Resilience training in the workplace: Developing skills for adaptability. Industrial and Organizational Psychology.

- Robertson, I. T., Cooper, C. L., Sarkar, M., & Curran, T. (2015). Resilience training in the workplace from 2003 to 2014: A systematic review. Journal of Occupational and Organizational Psychology.
- Robinson, K. (2015). Creative schools: The grassroots revolution that's transforming education. Viking.
- Rodrik, D. (2016). Premature deindustrialization. *Journal of Economic Growth.*
- Rogers, C. R. (1961). On becoming a person: A therapist's view of psychotherapy. Houghton Mifflin.
- Rogers, C. R. (2017). On becoming a person: A therapist's view of psychotherapy. Houghton Mifflin Harcourt.
- Romanosky, S. (2016). Examining the costs and causes of cyber incidents. Journal of Cybersecurity, 2(2), 121-135.
- Rometty, G. (2016). P-TECH: A New Model for Education. IBM Think Blog.
- Rometty, G. (2017). Preparing for the New Collar Jobs of the Future. IBM Think Blog.
- Roorda, D. L., Koomen, H. M., Spilt, J. L., & Oort, F. J. (2011). The influence of affective teacher-student relationships on students' school engagement and achievement: A meta-analytic approach. *Review of Educational Research, 81*(4), 493-529.
- Rosenblat, A., & Stark, L. (2016). Algorithmic Labor and Information Asymmetries: A Case Study of Uber's Drivers. International Journal of Communication.
- Rothwell, A., et al. (2019). Integrating Industry 4.0 into higher education institutions. Transnational Engineering Education using Advanced Technology.
- Russell, S., & Norvig, P. (2016). Artificial Intelligence: A Modern Approach. Pearson.
- Rüßmann, M., Lorenz, M., Gerbert, P., Waldner, M., Justus, J., Engel, P., & Harnisch, M. (2015). Industry 4.0: The future

of productivity and growth in manufacturing industries. Boston Consulting Group, 9(1), 54-89.

- Ryan, M., & Stahl, B. C. (2020). Artificial intelligence ethics guidelines for developers and users: Clarifying their content and normative implications. Journal of Information, Communication and Ethics in Society, 18(2), 159-178.
- Saez, E., & Zucman, G. (2019). The Triumph of Injustice: How the Rich Dodge Taxes and How to Make Them Pay. W. W. Norton & Company.
- Sale, K. (1995). *Rebels Against the Future: The Luddites and Their War on the Industrial Revolution: Lessons for the Computer Age.* Addison-Wesley.
- Sale, K. (1996). *Rebels Against the Future: The Luddites and Their War on the Industrial Revolution: Lessons for the Computer Age.* Addison-Wesley.
- Salesforce. (2014). Introducing Trailhead: A Fun Way to Learn Salesforce.
- Salovey, P., & Mayer, J. D. (1990). Emotional intelligence. Imagination, Cognition, and Personality, 9(3), 185-211.
- Savelsbergh, M., & Van Woensel, T. (2016). 50th Anniversary Invited Article—City Logistics: Challenges and Opportunities. Transportation Science, 50(2), 579-590.
- Sawyer, R. K. (2006). Educating for innovation. Thinking Skills and Creativity, 1(1), 41-48.
- Sawyer, R. K. (2007). Group genius: The creative power of collaboration. Basic Books.
- Schmidt, F. A. (2017). Digital labour markets in the platform economy: Mapping the political challenges of crowd work and gig work. Friedrich-Ebert-Stiftung.
- Schrank, D., Eisele, B., & Lomax, T. (2019). 2019 Urban Mobility Report. Texas A&M Transportation Institute.
- Schwab, K. (2016). The Fourth Industrial Revolution. World Economic Forum.

- Schwab, K. (2017). *The Fourth Industrial Revolution.* Crown Business.
- Schwab, K., & Davis, N. (2018). *Shaping the Fourth Industrial Revolution.* Currency.
- Schwab, K., & Samans, R. (2016). The Future of Jobs: Employment, Skills and Workforce Strategy for the Fourth Industrial Revolution. World Economic Forum.
- Schwartz, H., & Pellegrino, J. (2021). Preparing for the Future: The P-TECH Model and Policy Implications. Journal of Policy Analysis and Management.
- Schwartz, J., et al. (2019). *Deloitte Insights: The path to prosperity in the AI era.* Deloitte. Retrieved from [Deloitte website].
- Schwartz, J., et al. (2022). Workforce Reskilling: Corporate Strategies in the AI Era. Harvard Business Review.
- Schwartz, J., Hatfield, S., Jones, R., & Anderson, S. (2020). Work disrupted: Opportunity, resilience, and growth in the accelerated future of work. John Wiley & Sons.
- Selwyn, N. (2014). Digital technology and the contemporary university: Degrees of digitization. Routledge.
- Shane, S. (2003). A General Theory of Entrepreneurship: The Individual-Opportunity Nexus. Edward Elgar Publishing.
- Sharkey, N. (2010). Saying 'No!' to Lethal Autonomous Targeting. *Journal of Military Ethics.*
- Shaw, J., Rudzicz, F., Jamieson, T., & Goldfarb, A. (2018). Artificial intelligence and the implementation challenge. Journal of Medical Internet Research, 20(7), e10059.
- Shneiderman, B. (2016). The New ABCs of Research: Achieving Breakthrough Collaborations. Oxford University Press.
- Shute, V. J., & Zapata-Rivera, D. (2012). Adaptive Educational Systems. Adaptive Technologies for Training and Education. Cambridge University Press.
- Siciliano, B., & Khatib, O. (2016). *Springer Handbook of Robotics.* Springer.

- Siemens, G. (2005). Connectivism: A Learning Theory for the Digital Age. International Journal of Instructional Technology and Distance Learning.
- SkillsFuture Singapore. (2015). SkillsFuture: A pathway to future success.
- SkillsFuture Singapore. (2020). About SkillsFuture.
- Skovholt, T. M., & Trotter-Mathison, M. (2016). The resilient practitioner: Burnout and compassion fatigue prevention and self-care strategies for the helping professions. Routledge.
- Smith, A., & Anderson, J. (2014). AI, Robotics, and the Future of Jobs. Pew Research Center.
- Smith, J. (2018). The Value of Salesforce Certifications. Salesforce Certification Magazine.
- Smith, J. (2021). Aligning Employee Skills with Business Strategy: A Walmart Case Study. Retail Industry Analysis.
- Smith, J. (2021). The role of private sector in education and training: A review of Google's certificate programs. Journal of Vocational Education and Training.
- Smith, M. K. (2021). Cultivating a lifelong learning mindset: Strategies and implications for education. Journal of Lifelong Learning.
- Smith, R., & Leberstein, S. (2015). Rights on Demand: Ensuring Workplace Standards and Worker Security In the On-Demand Economy. National Employment Law Project.
- Sottilare, R. A., Graesser, A., Hu, X., & Olney, A. (2018). Design recommendations for adaptive intelligent tutoring systems: Adaptive instructional strategies. Army Research Laboratory.
- Sovacool, B. K., & Griffiths, S. (2020). The cultural barriers to renewable energy and energy efficiency in the United States. Nature Energy, 5(6), 462-467.
- Sparrow, R., & Sparrow, L. (2006). In the hands of machines? The future of aged care. *Minds and Machines*.
- Specht, K., Siebert, R., Hartmann, I., Freisinger, U. B., Sawicka, M., Werner, A., Thomaier, S., Henckel, D., Walk, H., &

Dierich, A. (2014). Urban agriculture of the future: an overview of sustainability aspects of food production in and on buildings. Agriculture and Human Values, 31(1), 33-51.

- Spencer, D. A. (2018). Fear and Hope in an Age of Mass Automation: Debating the Future of Work. New Technology, Work and Employment, 33(1), 1-12.
- Spencer, D. A., et al. (2020). Upskilling in the Digital Workforce: Training in a Time of Technological Change. Journal of Vocational Education & Training.
- Stahl, B. C. (2013). Responsible research and innovation: The role of privacy in an emerging framework. Science and Public Policy.
- Standing, G. (2017). Basic Income: And How We Can Make It Happen. Penguin Books.
- Stanton, P., & Thomas, R. (2018). Disruptive Innovation in Digital Platforms and the Rise of Competing Models of Employment. Employee Relations.
- Stephenson, R. (2016). Reskilling and Upskilling: The Key to Thriving in the Age of AI. Harvard Business Review.
- Stiglitz, J. E. (2019). People, Power, and Profits: Progressive Capitalism for an Age of Discontent. W. W. Norton & Company.
- Sue, D. W. (2001). Multidimensional Facets of Cultural Competence. The Counseling Psychologist.
- Sue, D. W., & Sue, D. (2012). Counseling the culturally diverse: Theory and practice. John Wiley & Sons.
- Sundararajan, A. (2016). The Sharing Economy: The End of Employment and the Rise of Crowd-Based Capitalism. MIT Press.
- Sundararajan, A. (2017). The Sharing Economy: The End of Employment and the Rise of Crowd-Based Capitalism. MIT Press.
- Susskind, R., & Susskind, D. (2015). The Future of the Professions: How Technology Will Transform the Work of Human Experts. *Oxford University Press.*

- Syverson, C. (2017). Challenges to mismeasurement explanations for the US productivity slowdown. Journal of Economic Perspectives, 31(2), 165-186.
- Taddeo, M. (2016). The ethical impact of data science. Current Opinion in Data Science, 1, 46-50.
- Taddeo, M., & Floridi, L. (2018). Regulate artificial intelligence to avert cyber arms race. *Nature*, 556(7701), 296-298.
- Taddy, M. (2019). Business Data Science: Combining Machine Learning and Economics to Optimize, Automate, and Accelerate Business Decisions. *McGraw-Hill Education.*
- Tan, O. S. (2019). Singapore's SkillsFuture initiative and its implications for higher education. In J. Shin & P. Teixeira (Eds.), Encyclopedia of International Higher Education Systems and Institutions. Dordrecht: Springer.
- Tegmark, M. (2017). *Life 3.0: Being Human in the Age of Artificial Intelligence.* Knopf.
- Tegmark, M. (2017). Life 3.0: Being Human in the Age of Artificial Intelligence. Knopf.
- Tetlock, P. E., & Gardner, D. (2015). *Superforecasting: The Art and Science of Prediction.* Random House.
- Tham, D. (2018). AI governance in Singapore: A model for other countries? *AI Ethics Journal,* 1(2), 37-46.
- The World Bank. (2020). Jobs in Renewable Energy Expanding Worldwide.
- Thomaier, S., Specht, K., Henckel, D., Dierich, A., Siebert, R., Freisinger, U. B., & Sawicka, M. (2015). Farming in and on urban buildings: Present practice and specific novelties of Zero-Acreage Farming (ZFarming). Renewable Agriculture and Food Systems, 30(1), 43-54.
- Thomas, J. W. (2000). A review of research on project-based learning. Autodesk Foundation.

- Ticona, J., & Mateescu, A. (2018). Trust and Safety in the Gig Economy: An Analysis of Worker Risk. Data & Society Research Institute.
- Tomlinson, C. A., & McTighe, J. (2006). Integrating differentiated instruction and understanding by design. ASCD.
- Topol, E. (2019). *Deep Medicine: How Artificial Intelligence Can Make Healthcare* Topol, E. J. (2019). High-performance medicine: the convergence of human and artificial intelligence. Nature Medicine, 25(1), 44-56.
- Topol, E. J. (2019). High-performance medicine: the convergence of human and artificial intelligence. Nature Medicine, 25(1), 44-56.
- Trilling, B., & Fadel, C. (2009). 21st Century Skills: Learning for Life in Our Times. Jossey-Bass.
- Trilling, B., & Fadel, C. (2018). 21st Century Skills: Learning for Life in Our Times. Jossey-Bass.
- Turkle, S. (2017). Alone Together: Why We Expect More from Technology and Less from Each Other. *Basic Books.*
- Twenge, J. M., & Campbell, W. K. (2018). Associations between screen time and lower psychological well-being among children and adolescents: Evidence from a population-based study. *Preventive Medicine Reports.*
- Twombly, S. B., Salisbury, M. H., Tumanut, S. D., & Klute, P. (2012). Study abroad in a new global century: Renewing the promise, refining the purpose. ASHE Higher Education Report.
- U.S. Bureau of Labor Statistics. (2020). Occupational Outlook Handbook: Solar Photovoltaic Installers.
- UNESCO. (2021). *Educational strategies to address AI challenges.* United Nations Educational, Scientific and Cultural Organization. Retrieved from [UNESCO website].
- United Nations. (2018). AI for Social Good. United Nations.
- University of Helsinki. (2019). Elements of AI: Bridging the Gap Between Academia and Industry.

- Uzzi, B. (1997). Social structure and competition in interfirm networks: The paradox of embeddedness. Administrative Science Quarterly, 42(1), 35-67.
- Vaillant, G. E., & Mukamal, K. (2001). Successful Aging. American Journal of Psychiatry.
- Valenduc, G., & Vendramin, P. (2017). Work in the Digital Economy: Sorting the Old from the New. ETUI Research Paper - Working Paper 2017.03.
- Van Dijk, J. (2020). The Digital Divide. *Polity Press.*
- Van Doorn, J., et al. (2017). The Impact of Customer Service on Customer Lifetime Value: An Empirical Investigation. Journal of Service Research.
- Van Huis, A., & Oonincx, D. G. (2017). The environmental sustainability of insects as food and feed. A review. Agronomy for Sustainable Development, 37(5), 43.
- Van Laar, E., et al. (2017). The relation between 21st-century skills and digital skills: A systematic literature review. Computers in Human Behavior.
- Van Laar, E., Van Deursen, A. J., Van Dijk, J. A., & De Haan, J. (2017). The relation between 21st-century skills and digital skills: A systematic literature review. Computers in Human Behavior, 72, 577-588.
- Van Parijs, P., & Vanderborght, Y. (2017). Basic Income: A Radical Proposal for a Free Society and a Sane Economy. Harvard University Press.
- Van Vlasselaer, V., Bravo, C., Caelen, O., Eliassi-Rad, T., Akoglu, L., Snoeck, M., & Baesens, B. (2015). APATE: A novel approach for automated credit card transaction fraud detection using network-based extensions. *Decision Support Systems*, 75, 38-48.
- Veletsianos, G., & Shepherdson, P. (2016). A systematic analysis and synthesis of the empirical MOOC literature published in 2013–2015. The International Review of Research in Open and Distributed Learning.

- Verhoef, P. C., Broekhuizen, T., Bart, Y., Bhattacharya, A., Dong, J., Fabian, N., & Haenlein, M. (2021). Digital Transformation: A Multidisciplinary Reflection and Research Agenda. Journal of Business Research, 122, 889-901.
- Verhoef, P. C., Stephen, A. T., Kannan, P. K., Luo, X., Abhishek, V., Andrews, M., ... & Zhang, Y. (2021). Consumer connectivity in a complex, technology-enabled, and mobile-oriented world with smart products. *Journal of Interactive Marketing*, 51, 1-8.
- Voigt, P., & Von dem Bussche, A. (2017). The EU General Data Protection Regulation (GDPR). Springer.
- Voogt, J., et al. (2015). Computational thinking in compulsory education: Towards an agenda for research and practice. Education and Information Technologies.
- Wagner, T. (2008). The Global Achievement Gap. Basic Books.
- Wagner, T., Brahm, T., & Kirchgeorg, M. (2020). Transforming the Higher Education System in Germany: Implications for the Integration of Digital Technologies in Teaching and Learning Environments. European Journal of Higher Education IT.
- Wajcman, J. (2017). Automation: Is it really different this time? British Journal of Sociology.
- Wall, D. S. (2007). Cybercrime: The transformation of crime in the information age. Polity.
- Wallach, W., & Allen, C. (2009). Moral Machines: Teaching Robots Right from Wrong. Oxford University Press.
- Walmart. (2018). Walmart Launches Live Better U Education Program.
- Walmart. (2019). Walmart's Live Better U Program.
- Walmart. (2020). Live Better U: Impact on Workforce Development.
- Walsh, T. (2018). Machines that think: The future of artificial intelligence. Prometheus Books.
- Wamba, S. F., Queiroz, M. M., Trinchera, L., & De Bourmont, M. (2020). Assessing the impact of artificial intelligence on

supply chain performance. The International Journal of Logistics Management.

- Warschauer, M., & Healey, D. (1998). Computers and language learning: An overview. Language Teaching.
- WEF. (2020). *World Economic Forum on Public-Private Partnerships in the Age of AI.* World Economic Forum. Retrieved from [WEF website].
- Wei, Y., Chan, S., Wang, Y., & Cai, M. (2019). Energy savings through artificial intelligence: A review. Renewable and Sustainable Energy Reviews.
- Weick, K. E., & Quinn, R. E. (1999). Organizational change and development. Annual Review of Psychology, 50, 361-386.
- West, D. M. (2018). The Future of Work: Robots, AI, and Automation. Brookings Institution Press.
- West, S. M., Whittaker, M., & Crawford, K. (2019). Discriminating Systems: Gender, Race, and Power in AI. AI Now Institute.
- Westerman, G., Bonnet, D., & McAfee, A. (2014). Leading digital: Turning technology into business transformation. Harvard Business Press.
- White, L. Jr. (1962). *Medieval Technology and Social Change.* Oxford University Press.
- Whittaker, M., Crawford, K., Dobbe, R., Fried, G., Kaziunas, E., Mathur, V., ... & Schwartz, O. (2018). AI Now Report 2018. AI Now Institute at New York University.
- Wilson, H. J., & Daugherty, P. R. (2018). Collaborative Intelligence: Humans and AI Are Joining Forces. *Harvard Business Review*, 96(4), 114-123.
- Wing, J. M. (2006). Computational thinking. Communications of the ACM.
- WIPO. (2020). *World Intellectual Property Report: The Geography of Innovation - Local Hotspots, Global Networks.* World Intellectual Property Organization. Retrieved from [WIPO website].

- Wiser, R., Bolinger, M., Hoen, B., Millstein, D., Rand, J., Barbose, G., ... & Seel, J. (2016). Wind Technologies Market Report. Lawrence Berkeley National Laboratory.
- Wood, A. J., Graham, M., Lehdonvirta, V., & Hjorth, I. (2019). Good Gig, Bad Gig: Autonomy and Algorithmic Control in the Global Gig Economy. Work, Employment and Society.
- World Bank. (2016). Digital dividends. World Development Report.
- World Bank. (2020). World Development Report 2020: Trading for Development in the Age of Global Value Chains. World Bank.
- World Economic Forum. (2018). The Future of Jobs Report 2018. *World Economic Forum.*
- World Economic Forum. (2018). Towards a Reskilling Revolution: A Future of Jobs for All.
- World Economic Forum. (2020). Jobs of Tomorrow: Mapping Opportunity in the New Economy.
- World Economic Forum. (2020). The Future of Jobs Report 2020. Geneva: World Economic Forum.
- World Health Organization. (2017). Digital health for the End TB Strategy: an agenda for action.
- Xie, H., Chu, H. C., Hwang, G. J., & Wang, C. C. (2019). Trends and Development of Learning Analytics and Big Data in Educational Technology. Interactive Learning Environments.
- Xie, H., Chu, H. C., Hwang, G. J., & Wang, C. C. (2019). Trends and development of learning analytics and their applications toward personalized learning: A systematic review and analysis of journal publications from 2007 to 2017. Computers & Education.
- Xie, H., et al. (2019). Adaptive learning based on individual learning styles in learning management systems. Journal of Educational Computing Research.
- Xu, D., et al. (2019). Predictive analytics in education: A comparison of deep learning frameworks. Computers & Education.

- Yalom, I. D. (2002). The gift of therapy: An open letter to a new generation of therapists and their patients. HarperCollins.
- Yang, A. (2018). The war on normal people: The truth about America's disappearing jobs and why universal basic income is our future. Hachette UK.
- Yorke, M. (2006). Employability in higher education: What it is – What it is not. The Higher Education Academy.
- Yukl, G. (2009). Leadership in organizations. Pearson Education.
- Yukl, G. (2013). Leadership in organizations. Pearson.
- Zawacki-Richter, O., et al. (2019). Systematic review of research on artificial intelligence applications in higher education – where are the educators? International Journal of Educational Technology in Higher Education.
- Zhang, L., Zhou, W., & Li, P. (2018). A review of current progress of wind energy prediction and power forecasting methods in the world. Renewable and Sustainable Energy Reviews, 90, 104-115.
- Zhao, F., Liu, R., & Zhang, L. (2019). Artificial Intelligence in Urban Traffic Management: Frameworks and Applications. Transportation Research Record, 2673(6), 35-45.
- Zhao, Y. (2019). Culturally relevant pedagogy: Clashing with the global educational reform movement. Teachers College Press.
- Zhao, Y. (2019). What Works May Hurt: Side Effects in Education. Teachers College Press.
- Zhao, Y., Zhang, G., Yang, Y., Huang, W., & Philip, S. Y. (2019). A Survey of Deep Learning-Driven Object Detection. ACM Computing Surveys.
- Zhavoronkov, A., Ivanenkov, Y. A., Aliper, A., Veselov, M. S., Aladinskiy, V. A., Aladinskaya, A. V., ... & Mamoshina, P. (2019). Deep learning enables rapid identification of potent DDR1 kinase inhibitors. *Nature Biotechnology*, 37(9), 1038-1040.
- Zheng, L., & Walsh, M. (2019). Designing for Flexibility in Learning in the Age of AI. Journal of Learning Design.

- Zhou, K., Liu, T., & Zhou, L. (2018). Industry 4.0: Towards future industrial opportunities and challenges. *Fuzzy Systems and Knowledge Discovery (FSKD)*, 12, 214-228.
- Zhou, L., Pan, S., Wang, J., & Vasilakos, A. V. (2020). Machine learning on big data: Opportunities and challenges. Neurocomputing, 237, 350-361.
- Zhou, M., Xu, W., Li, Z., & Yeung, D. S. (2020). Artificial Intelligence for Education: Knowledge and its Assessment in AI-Enabled Learning Ecosystems. Artificial Intelligence Review.
- Zhu, F., & Liu, Q. (2018). Competing with Complementors: An Empirical Look at Amazon.com. Strategic Management Journal.
- Zhu, M., Sari, A. R., & Lee, M. (2020). A systematic review of research on adaptive learning in educational settings. Educational Research Review.
- Zimmerman, B. J. (2013). From cognitive modeling to self-regulation: A social cognitive career path. Educational Psychologist.
- Zou, J., & Schiebinger, L. (2018). AI can be sexist and racist — it's time to make it fair. Nature, 559(7714), 324-326.
- Zuboff, S. (2019). The Age of Surveillance Capitalism: The Fight for a Human Future at the New Frontier of Power. *PublicAffairs*.